STRAIGHT
FORWARD
MATH

Basic Computational Skills

by S. Harold Collins

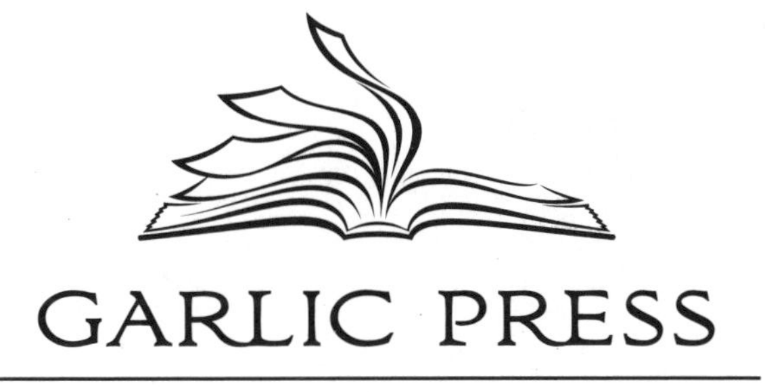

GARLIC PRESS

Educational Materials for Teachers and Parents

Printed in USA

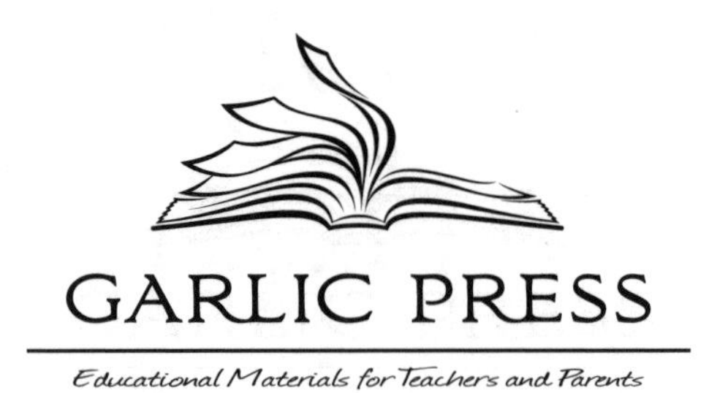

899 South College Mall Road
Bloomington, IN 47401

www.garlicpress.com

Publisher: Douglas M. Rife
Author: S. Harold Collins
Interior Design: Kathy Kifer
Cover Design: Kathy Kifer

To Parents and Teacher:

OVERVIEW

The *Straightforward Basic Computation Skills* has been designed for parents and teachers of children. It is a simple, straight forward way to increase computational fluency using the basic operations of addition, subtraction, multiplication, and division.

The skills gained in these pages parallel those recognized by local, state, and national entities seeking to strengthen the **standards** for mathematical **competence** and **performance**. More advanced computation is continued in our companion publication *Straightforward Advanced Computation Skills.*

Here are the components in the operations (+, –, x, ÷) to follow.

Beginning Assessment Test. The Beginning Assessment Test is arranged with 100 facts that will provide a beginning assessment for each operation. Notice that each Beginning Assessment Test has a special diagonal arrangement (See Answers, pages 79, 82, 85, and 88.) Each diagonal has a specific focus.

Practice Sheets. Practice Sheets are exercises that follow each Beginning Assessment Test. Start with a Practice Sheet that reflects information gained from the Beginning Assessment Test. If errors begin at a certain level as reflected in the Beginning Assessment Test, begin with a Practice Sheet at that level. Build up from there. Set standards to move to the next level.

Practice sheets are arranged in two forms: $\begin{array}{r} 4 \\ +9 \\ \hline \end{array}$ $\begin{array}{r} 9 \\ -4 \\ \hline \end{array}$ $\begin{array}{r} 7 \\ \times 3 \\ \hline \end{array}$ $4\overline{)12}$ and **4 + 4 =** **9 – 4 =** **7 x 3 =** **12 ÷ 4 =** . Both forms occur commonly.

Section Diagnostic Test. Periodically, a Section Diagnostic Test is presented. A Section Diagnostic Test is a good measure of skill attainment to that point. Section Diagnostic Tests are arranged to identify problems that may still exist within particular operations.

Set a standard to move from one section to the next. If the standard is not met, go back and focus on the problem area(s) with the Practice Sheets or similar materials.

Review Sheets. The Review Sheets are a preparation for the Final Assessment Test.

Final Assessment Test. The Final Assessment Test measures skill attainment. Compare the Final Assessment Test with the initial Beginning Assessment Test.

Contents

Addition

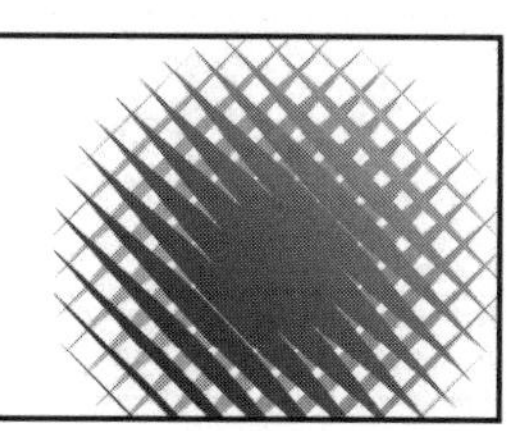

The **Beginning Assessment Test** has 100 problems, testing addition facts 0–10. The Beginning Assessment Test helps to establish a simple knowledge base. Answers on page 79 show the diagonal arrangement of facts.

After the Beginning Assessment Test, start **Practice Sheets** at an appropriate level as reflected by the Beginning Assessment Test results. If you can identify where errors start to occur–for instance, with 3s–start there. Or, simply begin at the lowest level (0s and 1s) and work up.

Some Practice Sheets provides problems in vertical form.

$$\begin{array}{r} 2 \\ +\ 6 \\ \hline \end{array} \qquad \begin{array}{r} 9 \\ +\ 7 \\ \hline \end{array} \qquad \begin{array}{r} 1 \\ +\ 8 \\ \hline \end{array}$$

The other Practice Sheets provide problems in horizontal form. Both forms are common.

8 + 7 =   **1 + 4 =**   **4 + 10 =**

A periodic **Section Diagnostic Test** measures skill attainment to that point. You can make a judgement after each Section Diagnostic Test: Have skills within that section been mastered? Set a standard to move on to the next section. If that standard is not met, go back and focus on the problem areas(s) with Practice Sheets, or with similar materials that you prepare.

There are two **Review Sheets**. The first Review Sheet is straightforward. The second Review Sheet is different. Problems may require answers, as in 5 + 6 = (sum). Or, the Review Sheet may require an addend that provides an answer.

5 + 6 = ___	**___ + 4 = 8**	**3 + ___ = 12**
Addend + Addend = Sum	**Addend + Addend = Sum**	**Addend + Addend = Sum**

The **Final Assessment Test** provides 100 problems in a straightforward form. Again, the facts are arranged diagonally (see Answers, page 81). With 100 problems, each problem is equal to 1%.

Beginning Assessment Test

$0 + 9$ $9 + 8$ $6 + 7$ $5 + 4$ $1 + 3$ $0 + 2$ $2 + 1$ $9 + 0$ $3 + 0$ $5 + 10$

$9 + 6$ $1 + 9$ $7 + 8$ $3 + 7$ $9 + 4$ $4 + 3$ $5 + 2$ $4 + 1$ $8 + 0$ $1 + 0$

$4 + 5$ $8 + 6$ $2 + 9$ $6 + 8$ $1 + 7$ $4 + 4$ $5 + 3$ $3 + 2$ $8 + 1$ $7 + 0$

$10 + 4$ $2 + 5$ $7 + 6$ $3 + 9$ $5 + 8$ $8 + 7$ $7 + 4$ $6 + 3$ $6 + 2$ $3 + 1$

$0 + 0$ $2 + 4$ $9 + 5$ $10 + 6$ $4 + 9$ $4 + 8$ $2 + 7$ $8 + 4$ $2 + 3$ $7 + 2$

$8 + 8$ $3 + 3$ $8 + 4$ $6 + 5$ $4 + 6$ $5 + 9$ $3 + 8$ $9 + 7$ $5 + 4$ $8 + 3$

$7 + 3$ $1 + 1$ $6 + 6$ $7 + 4$ $3 + 5$ $3 + 6$ $6 + 9$ $2 + 8$ $10 + 7$ $2 + 4$

$8 + 2$ $5 + 3$ $2 + 2$ $9 + 9$ $3 + 4$ $8 + 5$ $2 + 6$ $7 + 9$ $1 + 8$ $5 + 7$

$0 + 10$ $9 + 2$ $10 + 3$ $5 + 5$ $10 + 10$ $1 + 4$ $1 + 5$ $1 + 6$ $8 + 9$ $0 + 8$

$9 + 10$ $8 + 10$ $10 + 2$ $0 + 3$ $4 + 4$ $7 + 7$ $0 + 4$ $7 + 5$ $0 + 6$ $9 + 9$

Adding 0 & 1

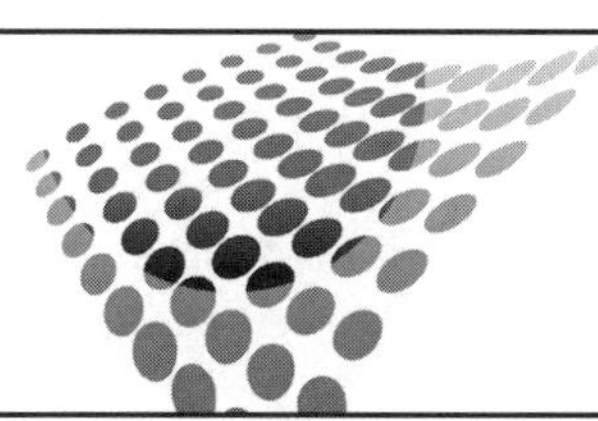

7 + 1	4 + 1	3 + 1	4 + 0	10 + 1	2 + 0	6 + 0	8 + 0	10 + 1
1 + 1	7 + 1	2 + 1	5 + 0	6 + 0	8 + 0	0 + 0	7 + 0	6 + 1
0 + 1	6 + 1	10 + 0	8 + 0	6 + 1	2 + 0	0 + 0	2 + 0	7 + 1
10 + 1	4 + 0	1 + 1	0 + 0	10 + 1	2 + 1	4 + 0	10 + 0	4 + 0
3 + 1	6 + 0	8 + 1	3 + 0	8 + 0	3 + 1	1 + 0	9 + 0	9 + 1
4 + 1	9 + 1	6 + 0	1 + 0	6 + 0	5 + 0	0 + 0	2 + 0	0 + 1
1 + 1	2 + 1	5 + 0	10 + 0	3 + 0	2 + 0	3 + 0	4 + 1	9 + 1
1 + 0	8 + 1	7 + 0	8 + 1	9 + 0	1 + 0	9 + 0	0 + 1	5 + 1
9 + 1	4 + 1	9 + 0	3 + 0	4 + 0	7 + 0	10 + 0	7 + 1	1 + 1

Adding 2

2 + 2 =	8 + 2 =	4 + 2 =	6 + 2 =	3 + 2 =
1 + 2 =	0 + 2 =	10 + 2 =	7 + 2 =	9 + 2 =
9 + 2 =	6 + 2 =	5 + 2 =	2 + 2 =	7 + 2 =
8 + 2 =	4 + 2 =	3 + 2 =	0 + 2 =	1 + 2 =
1 + 2 =	3 + 2 =	8 + 2 =	6 + 2 =	9 + 2 =
10 + 2 =	6 + 2 =	7 + 2 =	2 + 2 =	10 + 2 =
0 + 2 =	5 + 2 =	9 + 2 =	10 + 2 =	4 + 2 =
2 + 2 =	8 + 2 =	6 + 2 =	1 + 2 =	3 + 2 =
7 + 2 =	4 + 2 =	7 + 2 =	8 + 2 =	1 + 2 =
5 + 2 =	3 + 2 =	0 + 2 =	9 + 2 =	2 + 2 =
3 + 2 =	10 + 2 =	7 + 2 =	6 + 2 =	0 + 2 =
9 + 2 =	2 + 2 =	4 + 2 =	8 + 2 =	5 + 2 =
5 + 2 =	0 + 2 =	2 + 2 =	9 + 2 =	4 + 2 =
7 + 2 =	1 + 2 =	3 + 2 =	6 + 2 =	8 + 2 =
6 + 2 =	9 + 2 =	8 + 2 =	7 + 2 =	10 + 2 =
8 + 2 =	2 + 2 =	5 + 2 =	1 + 2 =	4 + 2 =
3 + 2 =	6 + 2 =	10 + 2 =	9 + 2 =	1 + 2 =
4 + 2 =	0 + 2 =	7 + 2 =	3 + 2 =	6 + 2 =

Adding 3

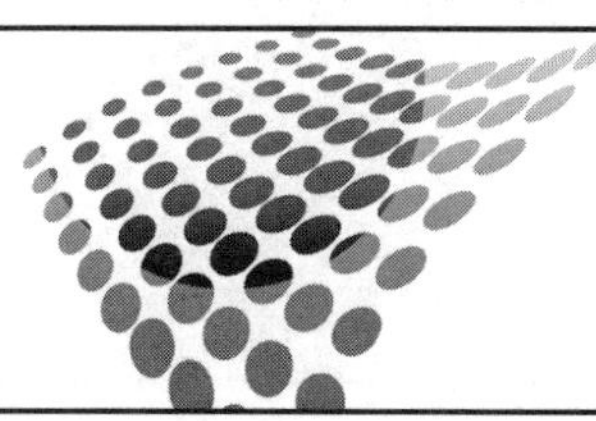

6 + 3	5 + 3	2 + 3	7 + 3	8 + 3	4 + 3	3 + 3	0 + 3	1 + 3
3 + 3	8 + 3	0 + 3	9 + 3	10 + 3	6 + 3	7 + 3	2 + 3	10 + 3
5 + 3	9 + 3	10 + 3	4 + 3	2 + 3	8 + 3	6 + 3	1 + 3	3 + 3
4 + 3	6 + 3	8 + 3	1 + 3	5 + 3	3 + 3	0 + 3	9 + 3	2 + 3
10 + 3	7 + 3	6 + 3	0 + 3	9 + 3	10 + 3	4 + 3	8 + 3	5 + 3
0 + 3	2 + 3	9 + 3	4 + 3	7 + 3	1 + 3	3 + 3	6 + 3	8 + 3
9 + 3	3 + 3	7 + 3	10 + 3	8 + 3	2 + 3	5 + 3	1 + 3	4 + 3
2 + 3	10 + 3	9 + 3	1 + 3	4 + 3	0 + 3	7 + 3	3 + 3	6 + 3
1 + 3	6 + 3	2 + 3	5 + 3	3 + 3	8 + 3	9 + 3	10 + 3	0 + 3

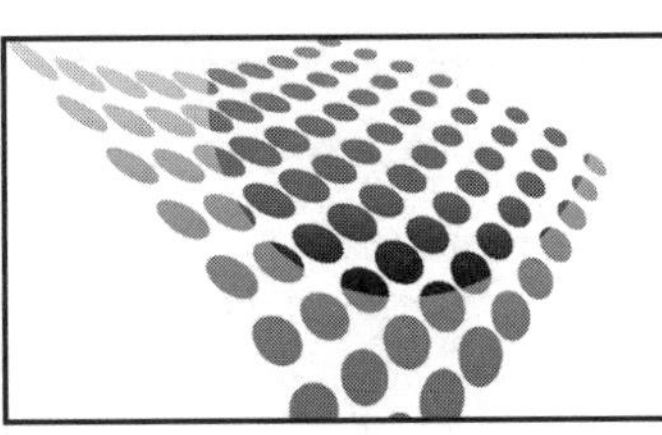

Adding 4

4 + 4 =	3 + 4 =	10 + 4 =	2 + 4 =	5 + 4 =
6 + 4 =	7 + 4 =	8 + 4 =	3 + 4 =	9 + 4 =
3 + 4 =	0 + 4 =	1 + 4 =	10 + 4 =	1 + 4 =
1 + 4 =	9 + 4 =	5 + 4 =	7 + 4 =	4 + 4 =
0 + 4 =	4 + 4 =	3 + 4 =	1 + 4 =	8 + 4 =
5 + 4 =	6 + 4 =	0 + 4 =	3 + 4 =	7 + 4 =
10 + 4 =	9 + 4 =	10 + 4 =	6 + 4 =	4 + 4 =
9 + 4 =	2 + 4 =	8 + 4 =	2 + 4 =	6 + 4 =
1 + 4 =	5 + 4 =	10 + 4 =	4 + 4 =	0 + 4 =
0 + 4 =	3 + 4 =	8 + 4 =	1 + 4 =	7 + 4 =
3 + 4 =	1 + 4 =	4 + 4 =	5 + 4 =	9 + 4 =
4 + 4 =	6 + 4 =	2 + 4 =	0 + 4 =	10 + 4 =
10 + 4 =	8 + 4 =	9 + 4 =	10 + 4 =	3 + 4 =
7 + 4 =	2 + 4 =	0 + 4 =	5 + 4 =	0 + 4 =
2 + 4 =	4 + 4 =	6 + 4 =	10 + 4 =	2 + 4 =
5 + 4 =	7 + 4 =	3 + 4 =	7 + 4 =	6 + 4 =
6 + 4 =	9 + 4 =	1 + 4 =	9 + 4 =	2 + 4 =
9 + 4 =	0 + 4 =	3 + 4 =	6 + 4 =	4 + 4 =

Adding 5

8 + 5	4 + 5	6 + 5	3 + 5	1 + 5	0 + 5	10 + 5	7 + 5	9 + 5
6 + 5	5 + 5	2 + 5	7 + 5	8 + 5	4 + 5	3 + 5	0 + 5	1 + 5
3 + 5	8 + 5	0 + 5	9 + 5	10 + 5	6 + 5	7 + 5	2 + 5	10 + 5
5 + 5	9 + 5	10 + 5	4 + 5	2 + 5	8 + 5	6 + 5	1 + 5	3 + 5
4 + 5	6 + 5	8 + 5	1 + 5	5 + 5	3 + 5	0 + 5	9 + 5	2 + 5
10 + 5	7 + 5	6 + 5	0 + 5	9 + 5	10 + 5	4 + 5	8 + 5	5 + 5
0 + 5	2 + 5	9 + 5	4 + 5	7 + 5	1 + 5	3 + 5	6 + 5	9 + 5
9 + 5	3 + 5	7 + 5	10 + 5	8 + 5	2 + 5	5 + 5	1 + 5	4 + 5
2 + 5	10 + 5	9 + 5	1 + 5	4 + 5	0 + 5	7 + 5	3 + 5	6 + 5

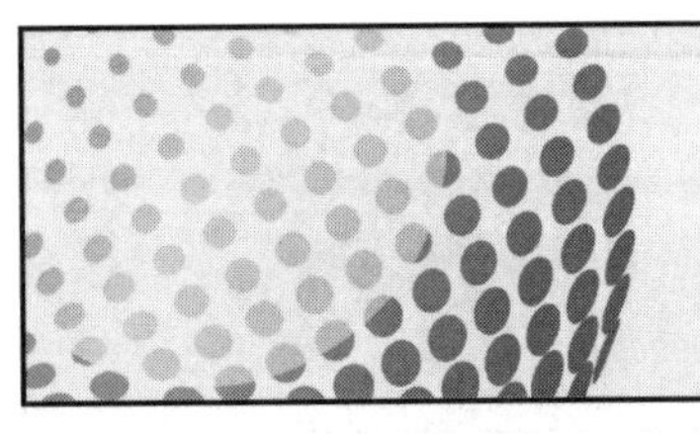

Section Diagnostic Test 0 – 5

0 + 5	5 + 4	2 + 3	2 + 2	4 + 1	1 + 0	5 + 2	3 + 3	4 + 4	5 + 5
2 + 5	0 + 4	5 + 3	3 + 2	3 + 1	4 + 0	1 + 2	5 + 3	3 + 4	4 + 5
4 + 5	2 + 4	0 + 3	5 + 2	2 + 1	3 + 0	4 + 2	1 + 3	5 + 4	3 + 5
3 + 5	4 + 4	2 + 3	0 + 2	5 + 1	2 + 0	3 + 2	4 + 3	1 + 4	5 + 5
5 + 5	3 + 4	4 + 3	2 + 2	0 + 1	5 + 0	2 + 2	3 + 3	4 + 4	1 + 5
1 + 5	5 + 4	3 + 3	4 + 2	2 + 1	0 + 0	5 + 2	2 + 3	3 + 4	4 + 5
4 + 5	1 + 4	5 + 3	3 + 2	4 + 1	2 + 0	0 + 2	5 + 3	2 + 4	3 + 5
3 + 5	4 + 4	1 + 3	5 + 2	3 + 1	4 + 0	2 + 2	0 + 3	5 + 4	2 + 5
2 + 5	3 + 4	3 + 3	1 + 2	5 + 1	3 + 0	4 + 2	2 + 3	0 + 4	5 + 5
5 + 5	2 + 4	4 + 3	4 + 2	1 + 1	5 + 0	3 + 2	4 + 3	2 + 4	0 + 5

Adding 6

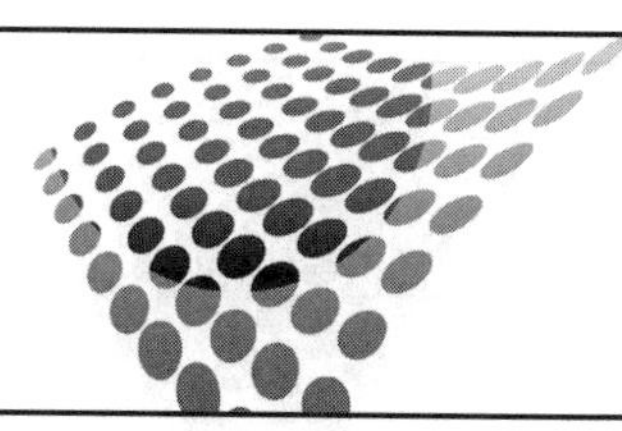

8 + 6 =	1 + 6 =	4 + 6 =	5 + 6 =	10 + 6 =
4 + 6 =	3 + 6 =	10 + 6 =	2 + 6 =	5 + 6 =
6 + 6 =	8 + 6 =	9 + 6 =	8 + 6 =	9 + 6 =
3 + 6 =	0 + 6 =	1 + 6 =	10 + 6 =	1 + 6 =
1 + 6 =	9 + 6 =	5 + 6 =	7 + 6 =	4 + 6 =
0+ 6 =	4 + 6 =	3 + 6 =	1 + 6 =	8 + 6 =
5 + 6 =	6 + 6 =	0 + 6 =	3 + 6 =	7 + 6 =
10 + 6 =	9 + 6 =	10 + 6 =	6 + 6 =	3 + 6 =
9 + 6 =	2 + 6 =	8 + 6 =	2 + 6 =	6 + 6 =
1 + 6 =	5 + 6 =	10 + 6 =	4 + 6 =	0 + 6 =
0 + 6 =	3 + 6 =	8 + 6 =	1 + 6 =	7 + 6 =
3 + 6 =	1 + 6 =	4 + 6 =	5 + 6 =	9 + 6 =
4 + 6 =	6 + 6 =	2 + 6 =	0 + 6 =	10 + 6 =
10 + 6 =	8 + 6 =	9 + 6 =	10 + 6 =	3 + 6 =
7 + 6 =	2 + 6 =	0 + 6 =	1 + 6 =	5 + 6 =
2 + 6 =	4 + 6 =	6 + 6 =	10 + 6 =	2 + 6 =
5 + 6 =	7 + 6 =	3 + 6 =	7 + 6 =	6 + 6 =
6 + 6 =	9 + 6 =	1 + 6 =	6 + 6 =	4 + 6 =

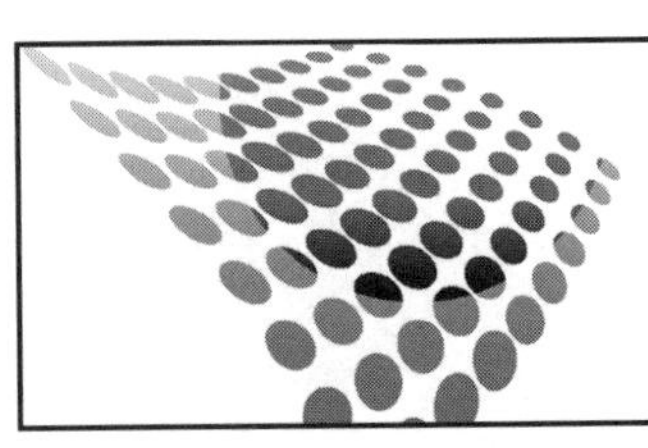

Adding 7

8 + 7	4 + 7	6 + 7	3 + 7	1 + 7	0 + 7	10 + 7	7 + 7	9 + 7
6 + 7	5 + 7	2 + 7	7 + 7	8 + 7	4 + 7	3 + 7	0 + 7	1 + 7
3 + 7	8 + 7	0 + 7	9 + 7	10 + 7	6 + 7	7 + 7	2 + 7	10 + 7
5 + 7	9 + 7	10 + 7	4 + 7	2 + 7	8 + 7	6 + 7	1 + 7	3 + 7
4 + 7	6 + 7	8 + 7	1 + 7	5 + 7	3 + 7	0 + 7	9 + 7	2 + 7
10 + 7	7 + 7	6 + 7	0 + 7	9 + 7	10 + 7	4 + 7	8 + 7	5 + 7
1 + 7	6 + 7	2 + 7	5 + 7	3 + 7	8 + 7	9 + 7	10 + 7	0 + 7
2 + 7	10 + 7	9 + 7	1 + 7	4 + 7	0 + 7	7 + 7	3 + 7	6 + 7
9 + 7	3 + 7	7 + 7	10 + 7	8 + 7	2 + 7	5 + 7	1 + 7	4 + 7

Adding 8

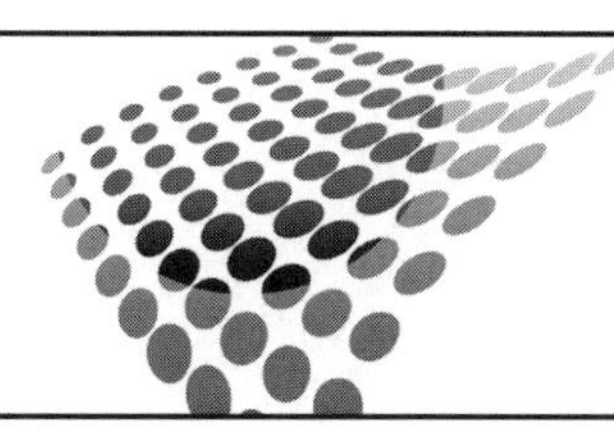

4 + 8 =	3 + 8 =	10 + 8 =	2 + 8 =	5 + 8 =
6 + 8 =	8 + 8 =	4 + 8 =	9 + 8 =	7 + 8 =
3 + 8 =	0 + 8 =	1 + 8 =	10 + 8 =	1 + 8 =
1 + 8 =	9 + 8 =	5 + 8 =	7 + 8 =	4 + 8 =
0 + 8 =	4 + 8 =	3 + 8 =	1 + 8 =	8 + 8 =
5 + 8 =	6 + 8 =	0 + 8 =	3 + 8 =	7 + 8 =
10 + 8 =	9 + 8 =	10 + 8 =	6 + 8 =	4 + 8 =
9 + 8 =	2 + 8 =	8 + 8 =	2 + 8 =	6 + 8 =
1 + 8 =	5 + 8 =	10 + 8 =	4 + 8 =	0 + 8 =
0 + 8 =	3 + 8 =	8 + 8 =	1 + 8 =	7 + 8 =
3 + 8 =	1 + 8 =	4 + 8 =	5 + 8 =	9 + 8 =
4 + 8 =	6 + 8 =	2 + 8 =	0 + 8 =	10 + 8 =
10 + 8 =	8 + 8 =	9 + 8 =	10 + 8 =	3 + 8 =
7 + 8 =	2 + 8 =	0 + 8 =	5 + 8 =	0 + 8 =
2 + 8 =	4 + 8 =	6 + 8 =	10 + 8 =	2 + 8 =
5 + 8 =	7 + 8 =	3 + 8 =	7 + 8 =	6 + 8 =
6 + 8 =	9 + 8 =	1 + 8 =	9 + 8 =	2 + 8 =
9 + 8 =	0 + 8 =	3 + 8 =	6 + 8 =	4 + 8 =

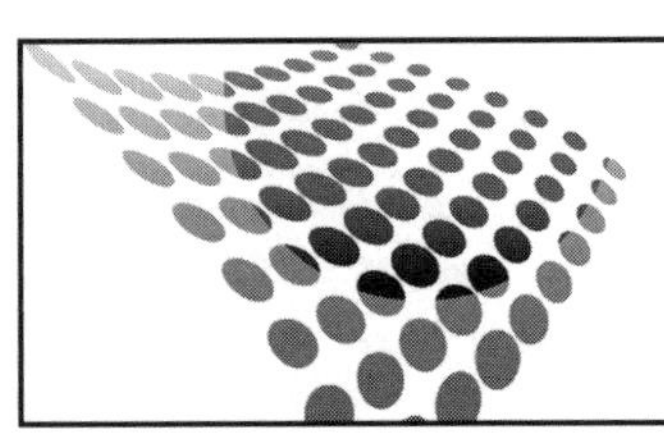

Adding 9

$8 + 9$ $4 + 9$ $6 + 9$ $3 + 9$ $1 + 9$ $0 + 9$ $10 + 9$ $7 + 9$ $9 + 9$

$6 + 9$ $5 + 9$ $2 + 9$ $7 + 9$ $8 + 9$ $4 + 9$ $3 + 9$ $0 + 9$ $1 + 9$

$3 + 9$ $8 + 9$ $0 + 9$ $9 + 9$ $10 + 9$ $6 + 9$ $7 + 9$ $2 + 9$ $10 + 9$

$5 + 9$ $9 + 9$ $10 + 9$ $4 + 9$ $2 + 9$ $8 + 9$ $6 + 9$ $1 + 9$ $3 + 9$

$4 + 9$ $6 + 9$ $8 + 9$ $1 + 9$ $5 + 9$ $3 + 9$ $0 + 9$ $9 + 9$ $2 + 9$

$10 + 9$ $7 + 9$ $6 + 9$ $0 + 9$ $9 + 9$ $10 + 9$ $4 + 9$ $8 + 9$ $5 + 9$

$0 + 9$ $2 + 9$ $9 + 9$ $4 + 9$ $7 + 9$ $1 + 9$ $3 + 9$ $6 + 9$ $9 + 9$

$9 + 9$ $3 + 9$ $7 + 9$ $10 + 9$ $8 + 9$ $2 + 9$ $5 + 9$ $1 + 9$ $4 + 9$

$2 + 9$ $10 + 9$ $9 + 9$ $1 + 9$ $4 + 9$ $0 + 9$ $7 + 9$ $3 + 9$ $6 + 9$

Adding 10

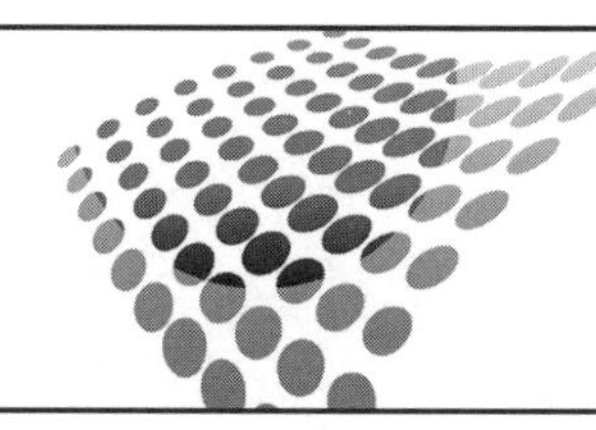

$8 + 10 =$	$1 + 10 =$	$4 + 10 =$	$5 + 10 =$	$10 + 10 =$
$4 + 10 =$	$3 + 10 =$	$10 + 10 =$	$2 + 10 =$	$5 + 10 =$
$6 + 10 =$	$8 + 10 =$	$9 + 10 =$	$8 + 10 =$	$9 + 10 =$
$3 + 10 =$	$0 + 10 =$	$1 + 10 =$	$10 + 10 =$	$1 + 10 =$
$1 + 10 =$	$9 + 10 =$	$5 + 10 =$	$7 + 10 =$	$4 + 10 =$
$0 + 10 =$	$4 + 10 =$	$3 + 10 =$	$1 + 10 =$	$8 + 10 =$
$5 + 10 =$	$6 + 10 =$	$0 + 10 =$	$3 + 10 =$	$7 + 10 =$
$10 + 10 =$	$9 + 10 =$	$10 + 10 =$	$6 + 10 =$	$3 + 10 =$
$9 + 10 =$	$2 + 10 =$	$8 + 10 =$	$2 + 10 =$	$6 + 10 =$
$1 + 10 =$	$5 + 10 =$	$10 + 10 =$	$4 + 10 =$	$0 + 10 =$
$0 + 10 =$	$3 + 10 =$	$8 + 10 =$	$1 + 10 =$	$7 + 10 =$
$3 + 10 =$	$1 + 10 =$	$4 + 10 =$	$5 + 10 =$	$9 + 10 =$
$4 + 10 =$	$6 + 10 =$	$2 + 10 =$	$0 + 10 =$	$10 + 10 =$
$10 + 10 =$	$8 + 10 =$	$9 + 10 =$	$10 + 10 =$	$3 + 10 =$
$7 + 10 =$	$2 + 10 =$	$0 + 10 =$	$1 + 10 =$	$5 + 10 =$
$2 + 10 =$	$4 + 10 =$	$6 + 10 =$	$10 + 10 =$	$2 + 10 =$
$5 + 10 =$	$7 + 10 =$	$3 + 10 =$	$7 + 10 =$	$6 + 10 =$
$6 + 10 =$	$9 + 10 =$	$1 + 10 =$	$6 + 10 =$	$4 + 10 =$

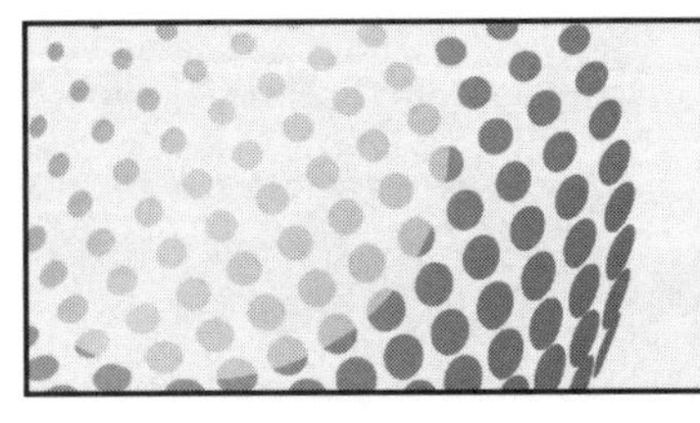

Section Diagnostic Test 6 – 10

2 + 6 | 9 + 7 | 1 + 8 | 7 + 9 | 3 +10 | 9 + 6 | 3 + 7 | 5 + 8 | 8 + 9 | 1 +10

4 + 6 | 8 + 7 | 3 + 8 | 6 + 9 | 5 +10 | 8 + 6 | 4 + 7 | 7 + 8 | 4 + 9 | 9 +10

6 + 6 | 10 + 7 | 9 + 8 | 5 + 9 | 7 +10 | 7 + 6 | 0 + 7 | 2 + 8 | 5 + 9 | 8 +10

8 + 6 | 6 + 7 | 0 + 8 | 4 + 9 | 9 +10 | 6 + 6 | 5 + 7 | 4 + 8 | 6 + 9 | 10 +10

10 + 6 | 5 + 7 | 6 + 8 | 3 + 9 | 10 +10 | 5 + 6 | 10 + 7 | 8 + 8 | 7 + 9 | 2 +10

1 + 6 | 5 + 7 | 10 + 8 | 2 + 9 | 8 +10 | 4 + 6 | 9 + 7 | 9 + 8 | 8 + 9 | 3 +10

3 + 6 | 4 + 7 | 7 + 8 | 1 + 9 | 6 +10 | 10 + 6 | 8 + 7 | 10 + 8 | 9 + 9 | 0 +10

5 + 6 | 2 + 7 | 5 + 8 | 0 + 9 | 4 +10 | 9 + 6 | 7 + 7 | 6 + 8 | 10 + 9 | 8 +10

7 + 6 | 1 + 7 | 8 + 8 | 10 + 9 | 2 +10 | 0 + 6 | 6 + 7 | 3 + 8 | 2 + 9 | 9 +10

4 + 6 | 3 + 7 | 1 + 8 | 2 + 9 | 6 +10 | 7 + 6 | 8 + 7 | 10 + 8 | 9 + 9 | 0 +10

Review Sheet

9 + 6	1 + 9	7 + 8	3 + 7	9 + 4	4 + 3	5 + 2	4 + 1	8 + 0
4 + 5	8 + 6	2 + 9	6 + 8	1 + 7	4 + 4	5 + 3	3 + 2	8 + 1
10 + 4	2 + 5	7 + 6	3 + 9	5 + 8	8 + 7	7 + 4	6 + 3	6 + 2
0 + 0	2 + 4	9 + 5	10 + 6	4 + 9	4 + 8	2 + 7	8 + 4	2 + 3
8 + 8	3 + 3	8 + 4	6 + 5	4 + 6	5 + 9	3 + 8	9 + 7	5 + 4
7 + 3	1 + 1	6 + 6	7 + 4	3 + 5	3 + 6	6 + 9	2 + 8	10 + 7
8 + 2	5 + 3	2 + 2	9 + 9	3 + 4	8 + 5	2 + 6	7 + 9	1 + 8
0 +10	9 + 2	10 + 3	5 + 5	10 +10	1 + 4	1 + 5	1 + 6	8 + 9
9 +10	8 +10	10 + 2	0 + 3	4 + 4	7 + 7	0 + 4	7 + 5	0 + 6

Review Sheet

5 + 6 = ___	___ + 4 = 8	3 + ___ = 12	___ + 7 = 9
6 + ___ = 7	5 + 7 = ___	___ + 3 = 7	3 + ___ = 11
___ + 7 = 10	6 + ___ = 8	5 + 8 = ___	___ + 2 = 6
8 + 8 = ___	___ + 4 = 11	6 + ___ = 9	5 + 9 = ___
9 + ___ = 17	8 + 9 = ___	___ + 5 = 12	6 + ___ = 10
___ + 9 = 18	9 + ___ = 19	8 + 1 = ___	___ + 6 = 13
9 + 5 = ___	___ + 3 = 12	9 + ___ = 15	8 + 2 = ___
8 + ___ = 12	4 + 9 = ___	___ + 2 = 9	9 + 4 = ___
___ + 6 = 14	8 + ___ = 17	9 + 8 = ___	___ + 5 = 15

9		8	3		4	5		8
+ 6	+ 9	+	+ 7	+ 4	+	+ 2	+ 1	+
	10	15		13	7		5	8

	7	2		1	4		3	8
+ 5	+	+ 9	+ 8	+	+ 4	+ 3	+	+ 1
9	15		14	8		8	5	

10	2		3	5		7	6	
+	+ 5	+ 6	+	+ 8	+ 7	+	+ 3	+ 2
14		13	12		13	11		8

0		9	10		4	2		2
+ 0	+ 4	+	+ 6	+ 9	+	+ 7	+ 4	+
	6	14		13	12		12	5

Final Assessment Test 0 – 10

$1 + 9$	$0 + 8$	$5 + 7$	$2 + 4$	$8 + 3$	$0 + 2$	$3 + 1$	$7 + 0$	$3 + 0$	$9 + 10$
$8 + 6$	$0 + 9$	$1 + 8$	$10 + 7$	$5 + 4$	$5 + 3$	$3 + 2$	$4 + 1$	$9 + 0$	$1 + 0$
$7 + 5$	$9 + 6$	$9 + 9$	$2 + 8$	$9 + 7$	$4 + 4$	$4 + 3$	$5 + 2$	$8 + 1$	$8 + 0$
$2 + 4$	$3 + 5$	$10 + 6$	$8 + 9$	$3 + 8$	$2 + 7$	$7 + 4$	$2 + 3$	$7 + 2$	$2 + 1$
$0 + 0$	$10 + 4$	$9 + 5$	$7 + 6$	$7 + 9$	$4 + 8$	$8 + 7$	$8 + 4$	$6 + 3$	$6 + 2$
$8 + 8$	$3 + 3$	$7 + 4$	$6 + 5$	$3 + 6$	$6 + 9$	$5 + 8$	$1 + 7$	$9 + 4$	$1 + 3$
$0 + 3$	$1 + 1$	$6 + 6$	$8 + 4$	$2 + 5$	$4 + 6$	$5 + 9$	$6 + 8$	$3 + 7$	$5 + 4$
$9 + 2$	$5 + 3$	$2 + 2$	$9 + 9$	$3 + 4$	$1 + 5$	$1 + 6$	$4 + 9$	$7 + 8$	$6 + 7$
$0 + 10$	$8 + 2$	$10 + 3$	$5 + 5$	$10 + 10$	$1 + 4$	$8 + 5$	$2 + 6$	$3 + 9$	$9 + 8$
$5 + 10$	$8 + 10$	$10 + 2$	$7 + 3$	$4 + 4$	$7 + 7$	$0 + 4$	$4 + 5$	$0 + 6$	$2 + 9$

Subtraction

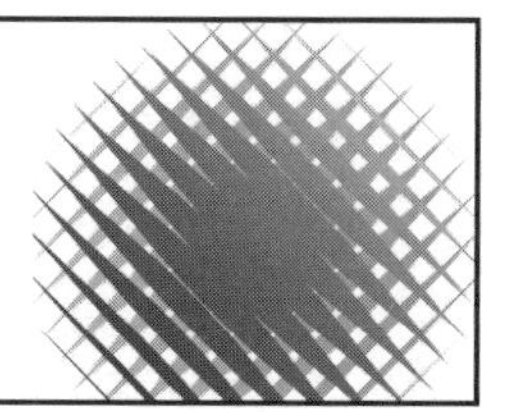

The **Beginning Assessment Test** has 100 problems, testing subtraction facts 0–10. The Beginning Assessment Test helps to establish a simple knowledge base. Answers on page 82 show the diagonal arrangement of facts.

After the Beginning Assessment Test, start **Practice Sheets** at an appropriate level as reflected by the Beginning Assessment Test results. If you can identify where errors start to occur–for instance, with 3s–start there. Or, simply begin at the lowest level (0s and 1s) and work up.

Some Practice Sheets provides problems in vertical form.

$$\begin{array}{r} 6 \\ -\ 2 \\ \hline \end{array} \qquad \begin{array}{r} 9 \\ -\ 7 \\ \hline \end{array} \qquad \begin{array}{r} 10 \\ -\ 4 \\ \hline \end{array}$$

The other Practice Sheets provide problems in horizontal form. Both forms are common.

6 – 2 = **9 – 7 =** **10 – 4 =**

A periodic **Section Diagnostic Test** measures skill attainment to that point. You can make a judgement after each Section Diagnostic Test: Have skills within that section been mastered? Set a standard to move on to the next section. If that standard is not met, go back and focus on the problem areas(s) with Practice Sheets, or with similar materials that you prepare.

There are two **Review Sheets**. The first Review Sheet is straightforward. The second Review Sheet has a difference. Problems may require answers, as in 6 – 2 = (difference). Or, the Review Sheet may require a minuend or subtrahend to provide an answer (difference).

6 – 2 = ___	**___ – 9 = 2**	**10 – ___ = 6**
Minuend – Subtrahend = Difference	**Minuend – Subtrahend = Difference**	**Minuend – Subtrahend = Difference**

The **Final Assessment Test** provides 100 problems in a straightforward form. Again, the facts are arranged diagonally (see Answers, page 84). With 100 problems, each problem is equal to 1%.

Beginning Assessment Test

$9 - 9$ $10 - 8$ $9 - 7$ $11 - 4$ $3 - 3$ $4 - 2$ $1 - 1$ $8 - 0$ $6 - 0$ $0 - 0$

$8 - 6$ $10 - 9$ $9 - 8$ $8 - 7$ $12 - 4$ $7 - 3$ $3 - 2$ $2 - 1$ $9 - 0$ $5 - 0$

$10 - 5$ $7 - 6$ $11 - 9$ $8 - 8$ $7 - 7$ $13 - 4$ $8 - 3$ $5 - 2$ $3 - 1$ $10 - 0$

$16 - 7$ $9 - 5$ $6 - 6$ $12 - 9$ $11 - 8$ $11 - 7$ $14 - 4$ $9 - 3$ $6 - 2$ $4 - 1$

$8 - 8$ $17 - 7$ $8 - 5$ $9 - 6$ $13 - 9$ $12 - 8$ $12 - 7$ $4 - 4$ $6 - 3$ $7 - 2$

$5 - 4$ $16 - 8$ $7 - 7$ $7 - 5$ $11 - 6$ $14 - 9$ $13 - 8$ $13 - 7$ $5 - 4$ $8 - 3$

$6 - 3$ $6 - 4$ $15 - 8$ $8 - 7$ $6 - 5$ $12 - 6$ $15 - 9$ $14 - 8$ $14 - 7$ $6 - 4$

$7 - 2$ $5 - 3$ $7 - 4$ $14 - 8$ $9 - 7$ $5 - 5$ $13 - 6$ $16 - 9$ $15 - 8$ $15 - 7$

$20 - 10$ $8 - 2$ $4 - 3$ $8 - 4$ $13 - 8$ $10 - 7$ $11 - 5$ $14 - 6$ $17 - 9$ $16 - 8$

$18 - 10$ $15 - 10$ $9 - 2$ $10 - 3$ $9 - 4$ $9 - 8$ $11 - 7$ $12 - 5$ $15 - 6$ $18 - 9$

Subtracting 0 & 1

$9 - 0$	$3 - 1$	$2 - 1$	$0 - 0$	$7 - 1$	$8 - 0$	$6 - 0$	$5 - 0$	$10 - 1$
$1 - 0$	$1 - 1$	$10 - 0$	$9 - 1$	$7 - 0$	$5 - 1$	$3 - 0$	$1 - 1$	$2 - 0$
$9 - 1$	$3 - 0$	$2 - 1$	$0 - 0$	$7 - 1$	$8 - 0$	$6 - 1$	$5 - 0$	$10 - 1$
$6 - 0$	$1 - 0$	$5 - 0$	$4 - 1$	$1 - 0$	$1 - 1$	$9 - 0$	$7 - 1$	$7 - 0$
$3 - 1$	$8 - 0$	$2 - 1$	$7 - 0$	$6 - 1$	$2 - 0$	$1 - 1$	$10 - 0$	$8 - 1$
$5 - 0$	$5 - 1$	$10 - 0$	$4 - 1$	$9 - 0$	$8 - 1$	$3 - 0$	$2 - 1$	$0 - 0$
$4 - 1$	$6 - 0$	$7 - 1$	$0 - 0$	$6 - 1$	$0 - 0$	$10 - 1$	$4 - 0$	$3 - 1$
$0 - 0$	$2 - 1$	$7 - 0$	$9 - 1$	$1 - 0$	$8 - 1$	$2 - 0$	$1 - 1$	$5 - 0$
$7 - 1$	$2 - 0$	$1 - 1$	$8 - 0$	$1 - 1$	$2 - 0$	$10 - 1$	$4 - 0$	$3 - 1$

Subtracting 2

2 – 2 =	8 – 2 =	4 – 2 =	6 – 2 =	3 – 2 =
8 – 2 =	4 – 2 =	3 – 2 =	12 – 2 =	10 – 2 =
9 – 2 =	5 – 2 =	6 – 2 =	2 – 2 =	7 – 2 =
4 – 2 =	6 – 2 =	7 – 2 =	5 – 2 =	2 – 2 =
11 – 2 =	3 – 2 =	8 – 2 =	12 – 2 =	9 – 2 =
2 – 2 =	8 – 2 =	6 – 2 =	10 – 2 =	3 – 2 =
12 – 2 =	5 – 2 =	9 – 2 =	7 – 2 =	4 – 2 =
5 – 2 =	3 – 2 =	11 – 2 =	9 – 2 =	2 – 2 =
7 – 2 =	4 – 2 =	6 – 2 =	8 – 2 =	10 – 2 =
9 – 2 =	2 – 2 =	4 – 2 =	12 – 2 =	5 – 2 =
3 – 2 =	11 – 2 =	7 – 2 =	6 – 2 =	12 – 2 =
7 – 2 =	3 – 2 =	10 – 2 =	9 – 2 =	6 – 2 =
5 – 2 =	11 – 2 =	2 – 2 =	8 – 2 =	4 – 2 =
8 – 2 =	2 – 2 =	5 – 2 =	3 – 2 =	8 – 2 =
6 – 2 =	9 – 2 =	3 – 2 =	7 – 2 =	10 – 2 =
4 – 2 =	11 – 2 =	7 – 2 =	3 – 2 =	6 – 2 =
8 – 2 =	7 – 2 =	5 – 2 =	9 – 2 =	12 – 2 =
10 – 2 =	8 – 2 =	9 – 2 =	7 – 2 =	5 – 2 =

Subtracting 3

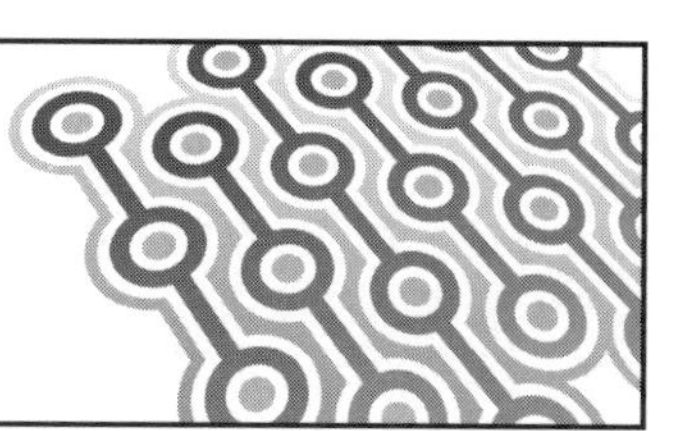

6	5	12	7	8	4	3	13	11
− 3	− 3	− 3	− 3	− 3	− 3	− 3	− 3	− 3
3	8	10	9	4	6	7	12	5
− 3	− 3	− 3	− 3	− 3	− 3	− 3	− 3	− 3
4	6	8	11	5	3	13	9	12
− 3	− 3	− 3	− 3	− 3	− 3	− 3	− 3	− 3
11	7	6	10	9	12	4	8	5
− 3	− 3	− 3	− 3	− 3	− 3	− 3	− 3	− 3
10	12	9	4	7	13	3	6	8
− 3	− 3	− 3	− 3	− 3	− 3	− 3	− 3	− 3
9	3	7	11	8	13	5	10	4
− 3	− 3	− 3	− 3	− 3	− 3	− 3	− 3	− 3
12	3	9	11	4	10	7	3	6
− 3	− 3	− 3	− 3	− 3	− 3	− 3	− 3	− 3
5	9	7	4	12	8	6	11	3
− 3	− 3	− 3	− 3	− 3	− 3	− 3	− 3	− 3
8	4	6	3	11	12	5	7	9
− 3	− 3	− 3	− 3	− 3	− 3	− 3	− 3	− 3

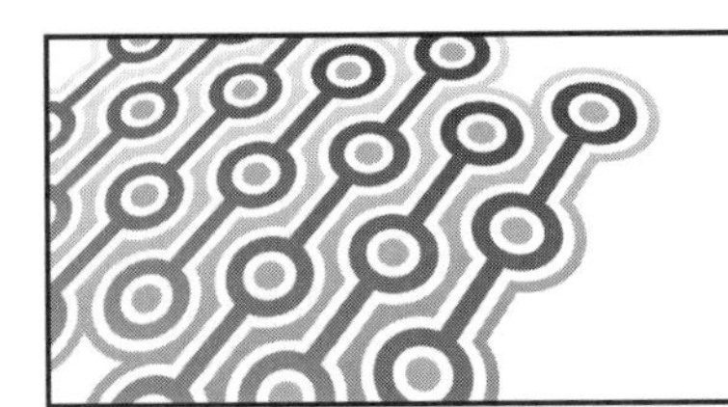

Subtracting 4

10 – 4 =	14 – 4 =	5 – 4 =	7 – 4 =	9 – 4 =
12 – 4 =	8 – 4 =	4 – 4 =	6 – 4 =	10 – 4 =
8 – 4 =	4 – 4 =	13 – 4 =	9 – 4 =	11 – 4 =
9 – 4 =	5 – 4 =	6 – 4 =	10 – 4 =	7 – 4 =
4 – 4 =	6 – 4 =	7 – 4 =	5 – 4 =	12 – 4 =
11 – 4 =	13 – 4 =	8 – 4 =	14 – 4 =	9 – 4 =
12 – 4 =	8 – 4 =	6 – 4 =	11 – 4 =	13 – 4 =
5 – 4 =	13 – 4 =	10 – 4 =	9 – 4 =	14 – 4 =
7 – 4 =	4 – 4 =	6 – 4 =	8 – 4 =	11 – 4 =
9 – 4 =	12 – 4 =	4 – 4 =	5 – 4 =	8 – 4 =
13 – 4 =	14 – 4 =	7 – 4 =	6 – 4 =	10 – 4 =
7 – 4 =	13 – 4 =	14 – 4 =	8 – 4 =	6 – 4 =
5 – 4 =	10 – 4 =	12 – 4 =	9 – 4 =	4 – 4 =
8 – 4 =	11 – 4 =	5 – 4 =	13 – 4 =	10 – 4 =
6 – 4 =	9 – 4 =	13 – 4 =	7 – 4 =	11 – 4 =
4 – 4 =	12 – 4 =	7 – 4 =	13 – 4 =	6 – 4 =
8 – 4 =	14 – 4 =	5 – 4 =	9 – 4 =	11 – 4 =
5 – 4 =	8 – 4 =	9 – 4 =	7 – 4 =	10 – 4 =

Subtracting 5

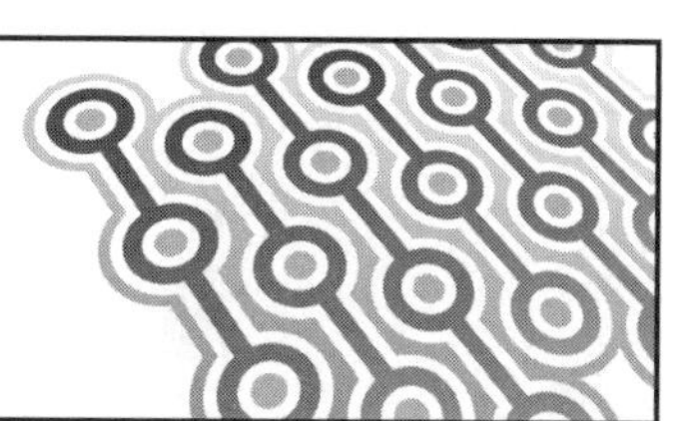

12	8	15	6	13	11	10	5	7
− 5	− 5	− 5	− 5	− 5	− 5	− 5	− 5	− 5
9	6	5	14	7	8	15	13	10
− 5	− 5	− 5	− 5	− 5	− 5	− 5	− 5	− 5
11	13	8	12	9	15	6	7	10
− 5	− 5	− 5	− 5	− 5	− 5	− 5	− 5	− 5
7	15	6	8	11	5	13	14	9
− 5	− 5	− 5	− 5	− 5	− 5	− 5	− 5	− 5
13	14	7	6	10	9	12	15	8
− 5	− 5	− 5	− 5	− 5	− 5	− 5	− 5	− 5
5	10	12	9	15	7	11	13	6
− 5	− 5	− 5	− 5	− 5	− 5	− 5	− 5	− 5
6	9	13	7	10	8	14	5	11
− 5	− 5	− 5	− 5	− 5	− 5	− 5	− 5	− 5
8	12	5	9	11	15	10	7	13
− 5	− 5	− 5	− 5	− 5	− 5	− 5	− 5	− 5
10	5	9	7	15	12	8	6	14
− 5	− 5	− 5	− 5	− 5	− 5	− 5	− 5	− 5

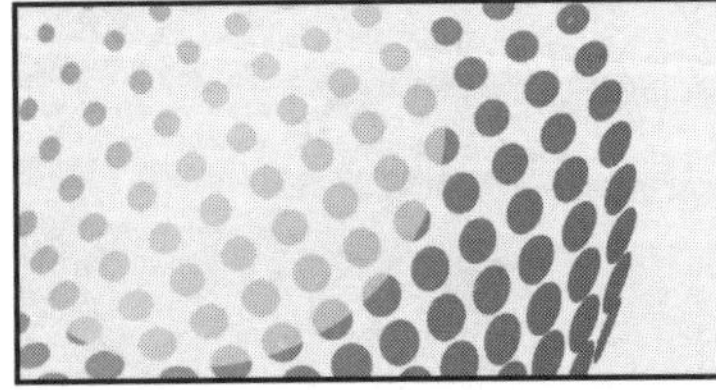

Section Diagnostic Test 0 – 5

5 – 5 =	11 – 5 =	7 – 5 =	9 – 5 =	8 – 5 =
9 – 4 =	6 – 4 =	10 – 4 =	13 – 4 =	7 – 4 =
7 – 3 =	11 – 3 =	5 – 3 =	6 – 3 =	12 – 3 =
10 – 2 =	2 – 2 =	5 – 2 =	8 – 2 =	11 – 2 =
1 – 1 =	9 – 1 =	6 – 1 =	10 – 1 =	7 – 1 =
0 – 0 =	2 – 0 =	10 – 0 =	7 – 0 =	6 – 0 =
6 – 5 =	12 – 5 =	14 – 5 =	11 – 5 =	8 – 5 =
12 – 4 =	9 – 4 =	7 – 4 =	14 – 4 =	12 – 4 =
4 – 3 =	13 – 3 =	10 – 3 =	12 – 3 =	9 – 3 =
9 – 2 =	7 – 2 =	4 – 2 =	6 – 2 =	10 – 2 =
3 – 1 =	5 – 1 =	8 – 1 =	7 – 1 =	1 – 1 =
5 – 0 =	1 – 0 =	9 – 0 =	8 – 0 =	0 – 0 =
10 – 5 =	8 – 5 =	12 – 5 =	15 – 5 =	9 – 5 =
8 – 4 =	5 – 4 =	7 – 4 =	10 – 4 =	6 – 4 =
9 – 3 =	10 – 3 =	8 – 3 =	11 – 3 =	9 – 3 =
8 – 2 =	3 – 2 =	10 – 2 =	12 – 2 =	2 – 2 =
4 – 4 =	14 – 4 =	11 – 4 =	8 – 4 =	7 – 4 =
12 – 5 =	9 – 5 =	10 – 5 =	15 – 5 =	13 – 5 =
4 – 4 =	13 – 4 =	9 – 4 =	6 – 4 =	10 – 4 =
6 – 3 =	4 – 3 =	12 – 3 =	13 – 3 =	3 – 3 =

Subtracting 6

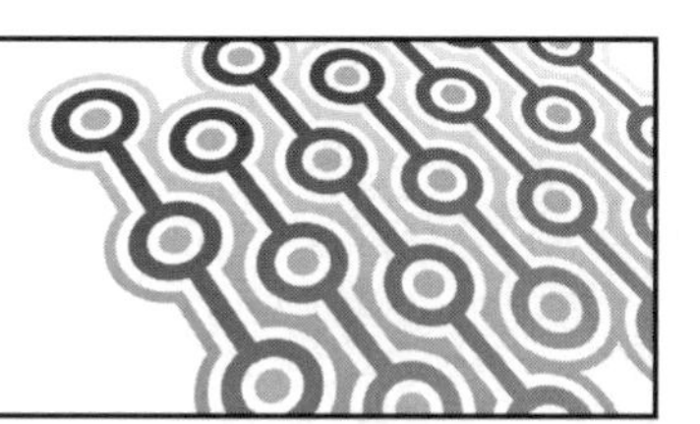

12 − 6	8 − 6	15 − 6	6 − 6	13 − 6	11 − 6	10 − 6	16 − 6	7 − 6
9 − 6	6 − 6	16 − 6	14 − 6	7 − 6	8 − 6	15 − 6	13 − 6	10 − 6
11 − 6	13 − 6	8 − 6	14 − 6	9 − 6	15 − 6	6 − 6	10 − 6	7 − 6
7 − 6	15 − 6	6 − 6	8 − 6	11 − 6	16 − 6	13 − 6	14 − 6	9 − 6
13 − 6	14 − 6	7 − 6	6 − 6	10 − 6	9 − 6	12 − 6	15 − 6	8 − 6
16 − 6	10 − 6	12 − 6	9 − 6	15 − 6	7 − 6	11 − 6	13 − 6	6 − 6
6 − 6	9 − 6	13 − 6	7 − 6	10 − 6	8 − 6	14 6	16 − 6	11 − 6
8 − 6	12 − 6	16 − 6	9 − 6	11 − 6	15 − 6	10 − 6	7 − 6	13 − 6
10 − 6	16 − 6	9 − 6	7 − 6	15 − 6	12 − 6	8 − 6	6 − 6	14 − 6

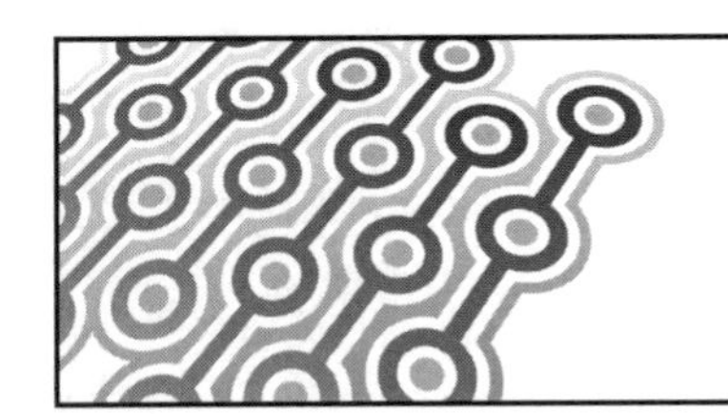

Subtracting 7

8 – 7 =	17 – 7 =	13 – 7 =	12 – 7 =	7 – 7 =
9 – 7 =	16 – 7 =	17 – 7 =	10 – 7 =	11 – 7 =
15 – 7 =	17 – 7 =	7 – 7 =	17 – 7 =	12 – 7 =
11 – 7 =	13 – 7 =	8 – 7 =	14 – 7 =	9 – 7 =
12 – 7 =	8 – 7 =	17 – 7 =	11 – 7 =	13 – 7 =
10 – 7 =	16 – 7 =	9 – 7 =	7 – 7 =	15 – 7 =
16 – 7 =	13 – 7 =	10 – 7 =	9 – 7 =	14 – 7 =
7 – 7 =	15 – 7 =	17 – 7 =	15 – 7 =	11 – 7 =
9 – 7 =	12 – 7 =	15 – 7 =	8 – 7 =	16 – 7 =
13 – 7 =	14 – 7 =	7 – 7 =	17 – 7 =	10 – 7 =
8 – 7 =	17 – 7 =	13 – 7 =	10 – 7 =	7 – 7 =
7 – 7 =	13 – 7 =	14 – 7 =	8 – 7 =	17 – 7 =
16 – 7 =	10 – 7 =	12 – 7 =	9 – 7 =	15 – 7 =
8 – 7 =	11 – 7 =	16 – 7 =	13 – 7 =	10 – 7 =
17 – 7 =	9 – 7 =	13 – 7 =	7 – 7 =	11 – 7 =
15 – 7 =	12 – 7 =	7 – 7 =	13 – 7 =	17 – 7 =
8 – 7 =	14 – 7 =	16 – 7 =	9 – 7 =	11 – 7 =
16 – 7 =	8 – 7 =	9 – 7 =	7 – 7 =	10 – 7 =

Subtracting 8

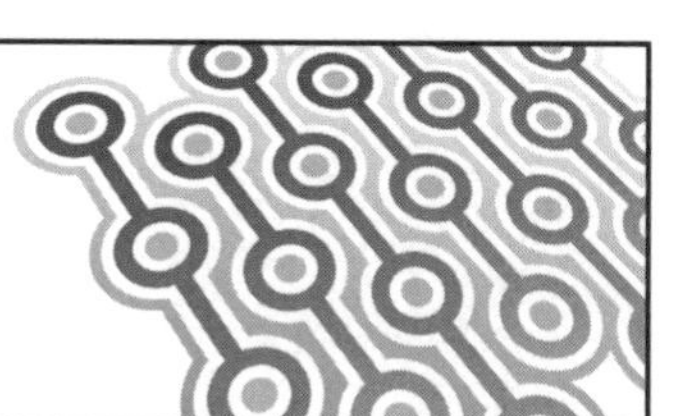

$8 - 8$ $15 - 8$ $17 - 8$ $13 - 8$ $11 - 8$ $10 - 8$ $16 - 8$ $18 - 8$ $9 - 8$

$17 - 8$ $16 - 8$ $14 - 8$ $18 - 8$ $8 - 8$ $15 - 8$ $13 - 8$ $10 - 8$ $11 - 8$

$13 - 8$ $8 - 8$ $14 - 8$ $9 - 8$ $15 - 8$ $17 - 8$ $18 - 8$ $10 - 8$ $16 - 8$

$15 - 8$ $17 - 8$ $8 - 8$ $11 - 8$ $16 - 8$ $13 - 8$ $14 - 8$ $9 - 8$ $12 - 8$

$14 - 8$ $18 - 8$ $17 - 8$ $10 - 8$ $9 - 8$ $12 - 8$ $15 - 8$ $8 - 8$ $15 - 8$

$10 - 8$ $12 - 8$ $9 - 8$ $15 - 8$ $18 - 8$ $11 - 8$ $13 - 8$ $17 - 8$ $8 - 8$

$9 - 8$ $13 - 8$ $18 - 8$ $10 - 8$ $8 - 8$ $14 - 8$ $16 - 8$ $11 - 8$ $15 - 8$

$12 - 8$ $16 - 8$ $9 - 8$ $11 - 8$ $15 - 8$ $10 - 8$ $18 - 8$ $13 - 8$ $18 - 8$

$16 - 8$ $9 - 8$ $18 - 8$ $15 - 8$ $12 - 8$ $17 - 8$ $9 - 8$ $14 - 8$ $13 - 8$

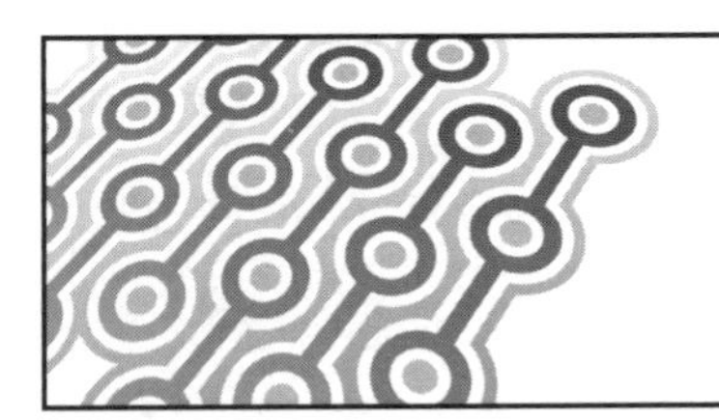

Subtracting 9

10 – 9 =	14 – 9 =	16 – 9 =	18 – 9 =	9 – 9 =
12 – 9 =	19 – 9 =	15 – 9 =	17 – 9 =	11 – 9 =
19 – 9 =	15 – 9 =	13 – 9 =	10 – 9 =	18 – 9 =
9 – 9 =	16 – 9 =	17 – 9 =	10 – 9 =	11 – 9 =
15 – 9 =	17 – 9 =	18 – 9 =	16 – 9 =	12 – 9 =
11 – 9 =	13 – 9 =	19 – 9 =	14 – 9 =	9 – 9 =
12 – 9 =	19 – 9 =	17 – 9 =	11 – 9 =	13 – 9 =
10 – 9 =	16 – 9 =	9 – 9 =	18 – 9 =	15 – 9 =
16 – 9 =	13 – 9 =	10 – 9 =	9 – 9 =	14 – 9 =
18 – 9 =	15 – 9 =	17 – 9 =	15 – 9 =	11 – 9 =
9 – 9 =	12 – 9 =	15 – 9 =	19 – 9 =	16 – 9 =
10 – 9 =	14 – 9 =	16 – 9 =	18 – 9 =	9 – 9 =
13 – 9 =	14 – 9 =	18 – 9 =	17 – 9 =	10 – 9 =
18 – 9 =	13 – 9 =	14 – 9 =	19 – 9 =	17 – 9 =
16 – 9 =	10 – 9 =	12 – 9 =	9 – 9 =	15 – 9 =
19 – 9 =	11 – 9 =	16 – 9 =	13 – 9 =	10 – 9 =
17 – 9 =	9 – 9 =	13 – 9 =	18 – 9 =	11 – 9 =
15 – 9 =	12 – 9 =	16 – 9 =	13 – 9 =	17 – 9 =

Subtracting 10

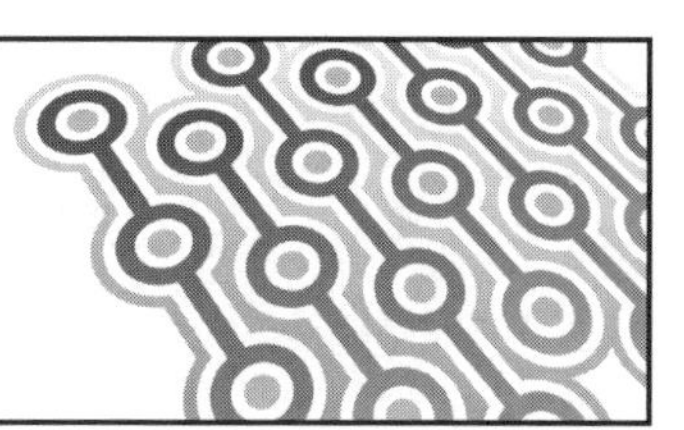

<table>
<tr><td>17
−10</td><td>16
−10</td><td>14
−10</td><td>18
−10</td><td>19
−10</td><td>15
−10</td><td>13
−10</td><td>10
−10</td><td>11
−10</td></tr>
<tr><td>13
−10</td><td>19
−10</td><td>14
−10</td><td>20
−10</td><td>10
−10</td><td>17
−10</td><td>18
−10</td><td>15
−10</td><td>16
−10</td></tr>
<tr><td>15
−10</td><td>17
−10</td><td>19
−10</td><td>11
−10</td><td>16
−10</td><td>13
−10</td><td>14
−10</td><td>15
−10</td><td>12
−10</td></tr>
<tr><td>14
−10</td><td>18
−10</td><td>17
−10</td><td>10
−10</td><td>20
−10</td><td>12
−10</td><td>15
−10</td><td>19
−10</td><td>15
−10</td></tr>
<tr><td>10
−10</td><td>12
−10</td><td>20
−10</td><td>15
−10</td><td>18
−10</td><td>11
−10</td><td>13
−10</td><td>17
−10</td><td>19
−10</td></tr>
<tr><td>20
−10</td><td>13
−10</td><td>18
−10</td><td>10
−10</td><td>19
−10</td><td>14
−10</td><td>16
−10</td><td>11
−10</td><td>15
−10</td></tr>
<tr><td>12
−10</td><td>16
−10</td><td>20
−10</td><td>11
−10</td><td>15
−10</td><td>10
−10</td><td>18
−10</td><td>13
−10</td><td>18
−10</td></tr>
<tr><td>16
−10</td><td>20
−10</td><td>18
−10</td><td>15
−10</td><td>12
−10</td><td>17
−10</td><td>20
−10</td><td>14
−10</td><td>13
−10</td></tr>
<tr><td>11
−10</td><td>17
−10</td><td>12
−10</td><td>16
−10</td><td>13
−10</td><td>18
−10</td><td>17
−10</td><td>18
−10</td><td>10
−10</td></tr>
</table>

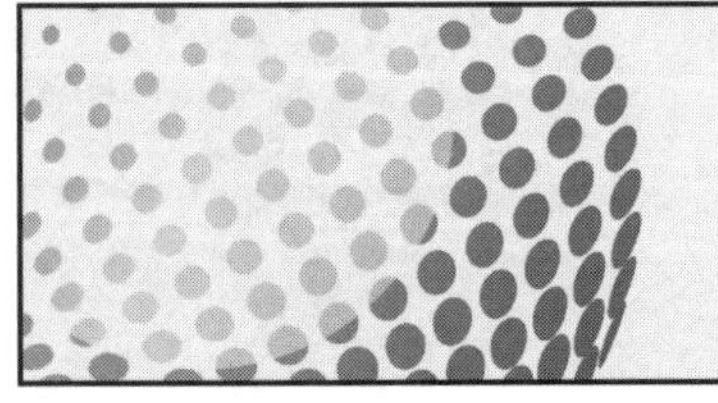

Section Diagnostic Test 6 – 10

9 – 6 =	7 – 7 =	18 – 8 =	15 – 9 =	15 – 10 =
10 – 6 =	17 – 7 =	12 – 8 =	17 – 9 =	10 – 10 =
12 – 6 =	8 – 7 =	17 – 8 =	16 – 9 =	17 – 10 =
9 – 6 =	16 – 7 =	11 – 8 =	12 – 9 =	12 – 10 =
11 – 6 =	15 – 7 =	10 – 8 =	9 – 9 =	18 – 10 =
7 – 6 =	10 – 7 =	16 – 8 =	10 – 9 =	14 – 10 =
10 – 6 =	14 – 7 =	9 – 8 =	13 – 9 =	19 – 10 =
8 – 6 =	11 – 7 =	17 – 8 =	18 – 9 =	13 – 10 =
13 – 6 =	13 – 7 =	8 – 8 =	14 – 9 =	11 – 10 =
15 – 6 =	9 – 7 =	15 – 8 =	11 – 9 =	16 – 10 =
12 – 6 =	12 – 7 =	11 – 8 =	19 – 9 =	20 – 10 =
6 – 6 =	7 – 7 =	13 – 8 =	9 – 9 =	12 – 10 =
10 – 6 =	8 – 7 =	15 – 8 =	13 – 9 =	15 – 10 =
14 – 6 =	16 – 7 =	10 – 8 =	15 – 9 =	14 – 10 =
11 – 6 =	9 – 7 =	14 – 8 =	17 – 9 =	11 – 10 =
16 – 6 =	10 – 7 =	16 – 8 =	12 – 9 =	16 – 10 =
9 – 6 =	15 – 7 =	8 – 8 =	14 – 9 =	17 – 10 =
15 – 6 =	13 – 7 =	18 – 8 =	11 – 9 =	13 – 10 =
13 – 6 =	11 – 7 =	12 – 8 =	10 – 9 =	10 – 10 =
16 – 6 =	14 – 7 =	9 – 8 =	16 – 9 =	18 – 10 =

Review Sheet

15 − 5	19 − 10	2 − 1	18 − 9	9 − 3	11 − 2	4 − 2	9 − 0	1 − 1
15 − 8	18 − 10	7 − 4	15 − 8	8 − 4	11 − 8	9 − 7	10 − 2	7 − 0
13 − 9	8 − 3	13 − 9	13 − 6	5 − 2	16 − 8	8 − 5	3 − 1	8 − 2
11 − 3	14 − 9	11 − 9	19 − 10	3 − 2	13 − 7	10 − 5	12 − 3	6 − 2
9 − 1	17 − 8	15 − 7	17 − 8	12 − 2	17 − 9	1 − 1	14 − 7	7 − 4
10 − 8	17 − 9	14 − 8	16 − 7	10 − 9	17 − 7	15 − 7	11 − 6	12 − 6
13 − 6	12 − 8	14 − 6	15 − 7	12 − 9	16 − 9	11 − 1	9 − 5	17 − 8
10 − 7	12 − 7	13 − 8	15 − 8	13 − 5	11 − 2	13 − 4	16 − 10	13 − 6
10 − 10	14 − 7	11 − 7	16 − 8	9 − 8	12 − 4	10 − 1	12 − 3	10 − 0

Review Sheet

16 – 8 = ___	___ – 6 = 1	11 – ___ = 2	___ – 8 = 0
13 – ___ = 9	8 – 3 = ___	___ – 2 = 3	10 – ___ = 10
___ – 7 = 9	9 – ___ = 4	12 – 9 = ___	___ – 8 = 3
11 – 7 = ___	___ – 4 = 10	9 – ___ = 6	6 – 2 = ___
17 – ___ = 10	8 – 5 = ___	___ – 6 = 3	13 – ___ = 4
___ – 8 = 5	12 – ___ = 5	6 – 3 = ___	___ – 2 = 5
16 – 8 = ___	___ – 7 = 0	7 – ___ = 2	11 – 6 = ___
13 – ___ = 5	13 – 7 = ___	___ – 4 = 1	8 – 3 = ___
___ – 9 = 5	13 – ___ = 6	15 – 7 = ___	___ – 7 = 1

6		15	14		6	7		7
– 5	– 6	–	– 8	– 7	–	– 2	– 3	–
	6	6		7	2		2	3

	13	16		15	20		13	10
– 8	–	– 9	– 8	–	–10	– 2	–	– 7
6	7		7	8		6	5	

11	6		16	10		15	9	
–	– 4	– 9	–	– 3	–10	–	– 4	– 7
6		8	8		8	5		4

0		17	13		11	17		9
– 0	–10	–	– 9	– 7	–	–10	– 9	–
	9	9		8	1		10	3

Final Assessment Test 0 – 10

$16 - 9$	$10 - 8$	$9 - 7$	$8 - 4$	$6 - 3$	$6 - 2$	$4 - 1$	$0 - 0$	$5 - 0$	$8 - 0$
$14 - 6$	$11 - 9$	$9 - 8$	$11 - 7$	$9 - 4$	$7 - 3$	$8 - 2$	$5 - 1$	$1 - 0$	$6 - 0$
$5 - 5$	$15 - 6$	$12 - 9$	$17 - 8$	$12 - 7$	$10 - 4$	$8 - 3$	$7 - 2$	$6 - 1$	$2 - 0$
$15 - 7$	$6 - 5$	$14 - 6$	$13 - 9$	$16 - 8$	$13 - 7$	$11 - 4$	$9 - 3$	$9 - 2$	$7 - 1$
$18 - 8$	$13 - 7$	$7 - 5$	$13 - 6$	$14 - 9$	$15 - 8$	$14 - 7$	$12 - 4$	$11 - 3$	$10 - 2$
$4 - 4$	$14 - 8$	$14 - 7$	$8 - 5$	$12 - 6$	$15 - 9$	$14 - 8$	$15 - 7$	$13 - 4$	$12 - 3$
$13 - 3$	$5 - 4$	$15 - 8$	$15 - 7$	$9 - 5$	$11 - 6$	$16 - 9$	$13 - 8$	$16 - 7$	$14 - 4$
$2 - 2$	$4 - 3$	$6 - 4$	$16 - 8$	$16 - 7$	$14 - 5$	$10 - 6$	$17 - 9$	$12 - 8$	$17 - 7$
$13 - 10$	$2 - 2$	$5 - 3$	$7 - 4$	$17 - 8$	$7 - 7$	$13 - 5$	$9 - 6$	$18 - 9$	$11 - 8$
$11 - 10$	$17 - 10$	$4 - 2$	$6 - 3$	$8 - 4$	$9 - 8$	$8 - 7$	$12 - 5$	$8 - 6$	$19 - 9$

Multiplication

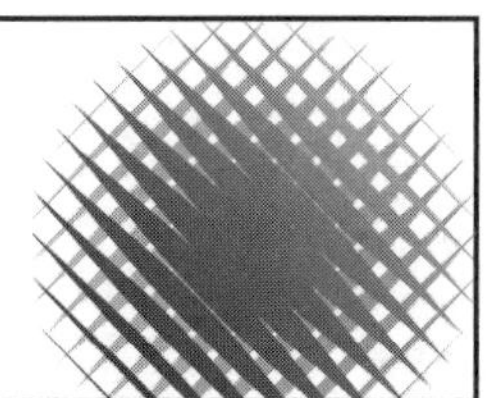

The **Beginning Assessment Test** has 100 problems, testing multiplication facts 0–10. The Beginning Assessment Test helps to establish a simple knowledge base. Answers on page 87 show the diagonal arrangement of facts.

After the Beginning Assessment Test, start **Practice Sheets** at an appropriate level as reflected by the Beginning Assessment Test results. If you can identify where errors start to occur–for instance, with 3s–start there. Or, simply begin at the lowest level (0s and 1s) and work up.

Some Practice Sheets provides problems in vertical form.

$$\begin{array}{r} 2 \\ \times\ 6 \\ \hline \end{array} \qquad \begin{array}{r} 9 \\ \times\ 7 \\ \hline \end{array} \qquad \begin{array}{r} 1 \\ \times\ 8 \\ \hline \end{array}$$

The other Practice Sheets provide problems in horizontal form. Both forms are common.

$6 \times 2 =$ $\quad$ $9 \times 7 =$ $\quad$ $1 \times 8 =$

A periodic **Section Diagnostic Test** measures skill attainment to that point. You can make a judgement after each Section Diagnostic Test: Have skills within that section been mastered? Set a standard to move on to the next section. If that standard is not met, go back and focus on the problem areas(s) with Practice Sheets, or with similar materials that you prepare.

There are two **Review Sheets**. The first Review Sheet is straightforward. The second Review Sheet is different. Problems may require answers, as in 6 x 2 = (product). Or, the Review Sheet may require factors that provide an answer (product).

$6 \times 2 =$ ___	___ $\times 7 = 63$	$1 \times$ ___ $= 8$
Factor x Factor = Product	Factor x Factor = Product	Factor x Factor = Product

The **Final Assessment Test** provides 100 problems in a straightforward form. Again, the facts are arranged diagonally (see Answers, page 87). With 100 problems, each problem is equal to 1%.

Beginning Assessment Test

0×9 9×8 4×7 8×5 8×6 5×2 5×1 9×0 3×0 4×10

1×6 1×9 8×8 9×7 5×5 4×6 8×2 2×1 5×0 8×0

3×4 7×3 2×9 0×8 3×7 6×5 9×6 7×2 3×1 6×0

9×2 5×4 3×3 3×9 3×8 7×7 4×5 5×6 6×2 4×1

2×2 4×2 8×4 4×3 4×9 7×8 2×7 3×5 7×6 9×2

8×8 7×7 0×2 4×4 6×3 5×9 2×8 5×7 9×5 3×6

9×1 1×1 3×3 3×2 9×4 5×3 6×9 4×8 6×7 7×5

6×10 6×1 5×5 6×6 6×2 6×4 2×3 7×9 6×8 8×7

7×10 0×10 7×1 9×9 4×4 7×2 2×4 9×3 8×9 5×8

2×10 5×10 9×10 8×1 0×0 5×5 5×2 7×4 8×3 9×9

Multiplying 0 & 1

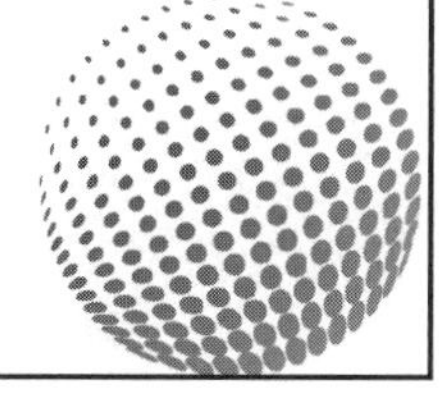

3×1 5×0 0×1 9×1 0×0 4×0 7×1 9×0 5×1

2×0 8×1 2×1 6×0 1×1 1×0 6×1 7×0 4×1

7×1 0×0 5×0 3×0 9×1 0×1 5×1 9×0 4×0

8×0 8×1 6×0 2×1 1×1 7×0 4×1 2×0 6×1

9×0 8×0 3×0 0×1 4×0 9×1 3×1 0×0 7×1

6×0 8×1 2×1 1×0 2×0 6×1 4×1 5×1 7×0

5×0 9×1 0×1 4×0 3×1 0×0 7×1 3×0 2×1

8×1 7×0 1×1 3×0 6×0 1×0 5×1 2×0 6×1

2×1 0×1 3×1 0×0 9×1 8×0 5×0 7×1 9×0

Multiplying 2

$2 \times 2 =$	$5 \times 2 =$	$4 \times 2 =$	$6 \times 2 =$	$3 \times 2 =$
$1 \times 2 =$	$0 \times 2 =$	$5 \times 2 =$	$7 \times 2 =$	$9 \times 2 =$
$9 \times 2 =$	$6 \times 2 =$	$8 \times 2 =$	$2 \times 2 =$	$7 \times 2 =$
$8 \times 2 =$	$4 \times 2 =$	$3 \times 2 =$	$5 \times 2 =$	$1 \times 2 =$
$1 \times 2 =$	$3 \times 2 =$	$8 \times 2 =$	$0 \times 2 =$	$9 \times 2 =$
$4 \times 2 =$	$6 \times 2 =$	$7 \times 2 =$	$2 \times 2 =$	$5 \times 2 =$
$0 \times 2 =$	$5 \times 2 =$	$9 \times 2 =$	$7 \times 2 =$	$4 \times 2 =$
$2 \times 2 =$	$8 \times 2 =$	$6 \times 2 =$	$1 \times 2 =$	$3 \times 2 =$
$7 \times 2 =$	$4 \times 2 =$	$5 \times 2 =$	$8 \times 2 =$	$1 \times 2 =$
$5 \times 2 =$	$3 \times 2 =$	$0 \times 2 =$	$9 \times 2 =$	$2 \times 2 =$
$3 \times 2 =$	$1 \times 2 =$	$7 \times 2 =$	$6 \times 2 =$	$0 \times 2 =$
$9 \times 2 =$	$2 \times 2 =$	$4 \times 2 =$	$8 \times 2 =$	$5 \times 2 =$
$5 \times 2 =$	$0 \times 2 =$	$3 \times 2 =$	$9 \times 2 =$	$4 \times 2 =$
$7 \times 2 =$	$1 \times 2 =$	$2 \times 2 =$	$6 \times 2 =$	$8 \times 2 =$
$6 \times 2 =$	$9 \times 2 =$	$3 \times 2 =$	$7 \times 2 =$	$0 \times 2 =$
$8 \times 2 =$	$2 \times 2 =$	$5 \times 2 =$	$1 \times 2 =$	$4 \times 2 =$
$5 \times 2 =$	$9 \times 2 =$	$8 \times 2 =$	$9 \times 2 =$	$7 \times 2 =$
$4 \times 2 =$	$0 \times 2 =$	$7 \times 2 =$	$3 \times 2 =$	$6 \times 2 =$

Multiplying 3

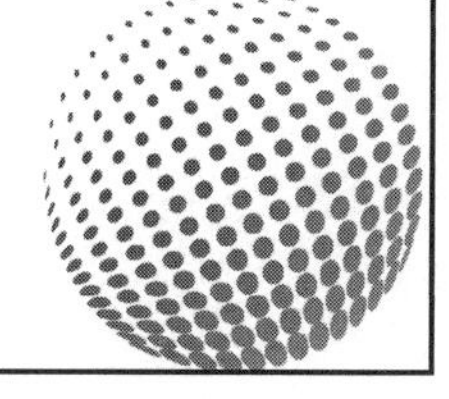

2×3 8×3 4×3 6×3 3×3 1×3 0×3 5×3 7×3

9×3 6×3 5×3 2×3 7×3 8×3 4×3 3×3 0×3

1×3 3×3 8×3 0×3 9×3 4×3 6×3 7×3 2×3

7×3 4×3 6×3 8×3 1×3 5×3 3×3 0×3 9×3

3×3 1×3 7×3 6×3 0×3 9×3 2×3 4×3 8×3

5×3 0×3 2×3 9×3 4×3 7×3 1×3 3×3 6×3

6×3 9×3 3×3 7×3 0×3 8×3 2×3 5×3 1×3

8×3 2×3 5×3 9×3 1×3 4×3 0×3 7×3 3×3

0×3 5×3 9×3 7×3 4×3 2×3 8×3 6×3 1×3

Multiplying 4

8 × 4 =	4 × 4 =	3 × 4 =	0 × 4 =	1 × 4 =
9 × 4 =	5 × 4 =	6 × 4 =	2 × 4 =	7 × 4 =
4 × 4 =	6 × 4 =	7 × 4 =	5 × 4 =	2 × 4 =
1 × 4 =	3 × 4 =	8 × 4 =	0 × 4 =	9 × 4 =
2 × 4 =	8 × 4 =	6 × 4 =	1 × 4 =	3 × 4 =
0 × 4 =	5 × 4 =	9 × 4 =	7 × 4 =	4 × 4 =
5 × 4 =	3 × 4 =	0 × 4 =	9 × 4 =	2 × 4 =
7 × 4 =	4 × 4 =	6 × 4 =	1 × 4 =	8 × 4 =
9 × 4 =	2 × 4 =	4 × 4 =	8 × 4 =	5 × 4 =
3 × 4 =	1 × 4 =	7 × 4 =	6 × 4 =	0 × 4 =
7 × 4 =	3 × 4 =	1 × 4 =	8 × 4 =	6 × 4 =
5 × 4 =	0 × 4 =	2 × 4 =	9 × 4 =	4 × 4 =
8 × 4 =	2 × 4 =	5 × 4 =	3 × 4 =	1 × 4 =
6 × 4 =	9 × 4 =	3 × 4 =	7 × 4 =	0 × 4 =
4 × 4 =	0 × 4 =	7 × 4 =	3 × 4 =	6 × 4 =
8 × 4 =	2 × 4 =	5 × 4 =	9 × 4 =	1 × 4 =
5 × 4 =	8 × 4 =	9 × 4 =	7 × 4 =	0 × 4 =
6 × 4 =	1 × 4 =	4 × 4 =	6 × 4 =	3 × 4 =

Multiplying 5

9 × 5   6 × 5   5 × 5   2 × 5   7 × 5   8 × 5   4 × 5   3 × 5   0 × 5

4 × 5   1 × 5   6 × 5   2 × 5   5 × 5   3 × 5   8 × 5   9 × 5   7 × 5

1 × 5   3 × 5   8 × 5   0 × 5   9 × 5   4 × 5   6 × 5   7 × 5   2 × 5

7 × 5   4 × 5   6 × 5   8 × 5   1 × 5   5 × 5   3 × 5   0 × 5   9 × 5

3 × 5   1 × 5   7 × 5   6 × 5   0 × 5   9 × 5   2 × 5   4 × 5   8 × 5

5 × 5   0 × 5   2 × 5   9 × 5   4 × 5   7 × 5   1 × 5   3 × 5   6 × 5

6 × 5   9 × 5   3 × 5   7 × 5   0 × 5   8 × 5   2 × 5   5 × 5   1 × 5

8 × 5   2 × 5   5 5   9 × 5   1 × 5   4 × 5   0 × 5   7 × 5   3 × 5

0 × 5   5 × 5   9 × 5   7 × 5   4 × 5   2 × 5   8 × 5   6 × 5   1 × 5

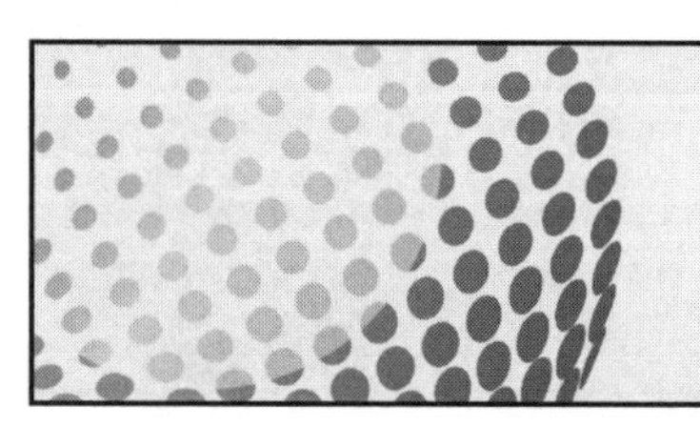

Section Diagnostic Test 0 – 5

4×5	5×4	5×3	7×2	8×1	5×0	6×2	3×3	3×4	3×5
5×5	6×4	9×3	2×2	5×1	4×0	1×2	0×3	6×4	5×5
7×5	9×4	2×3	0×2	3×1	6×0	9×2	5×3	7×4	7×5
9×5	8×4	1×3	5×2	3×1	7×0	8×2	1×3	3×4	9×5
8×5	0×4	3×3	4×2	4×1	7×0	3×2	8×3	0×4	1×5
2×5	1×4	5×3	2×2	6×1	5×0	3×2	5×3	8×4	9×5
4×5	9 4	7×3	4×2	9×1	5×0	5×2	8×3	4×4	7×5
6×5	3×4	9×3	8×2	6×1	6×0	7×2	6×3	5×4	8×5
8×5	5×4	3×3	0×2	4×1	7×0	6×2	6×3	7×4	5×5
0×5	2×4	0×3	9×2	0×1	5×0	1×2	9×3	6×4	1×5

Multiplying 6

8 × 6 =	4 × 6 =	3 × 6 =	0 × 6 =	1 × 6 =
9 × 6 =	5 × 6 =	6 × 6 =	2 × 6 =	7 × 6 =
4 × 6 =	6 × 6 =	7 × 6 =	5 × 6 =	2 × 6 =
1 × 6 =	3 × 6 =	8 × 6 =	0 × 6 =	9 × 6 =
2 × 6 =	8 × 6 =	6 × 6 =	1 × 6 =	3 × 6 =
0 × 6 =	5 × 6 =	9 × 6 =	7 × 4 =	4 × 6 =
5 × 6 =	3 × 6 =	0 × 6 =	9 × 6 =	2 × 6 =
7 × 6 =	4 × 6 =	6 × 6 =	8 × 6 =	1 × 6 =
9 × 6 =	2 × 6 =	4 × 6 =	3 × 6 =	5 × 6 =
3 × 6 =	1 × 6 =	7 × 6 =	6 × 6 =	0 × 6 =
7 × 6 =	3 × 6 =	1 × 6 =	8 × 6 =	6 × 6 =
5 × 6 =	0 × 6 =	2 × 6 =	9 × 6 =	4 × 6 =
8 × 6 =	2 × 6 =	5 × 6 =	3 × 6 =	1 × 6 =
6 × 6 =	9 × 6 =	3 × 6 =	7 × 6 =	0 × 6 =
4 × 6 =	0 × 6 =	7 × 6 =	3 × 6 =	6 × 6 =
8 × 6 =	2 × 6 =	5 × 6 =	9 × 6 =	4 × 6 =
5 × 6 =	8 × 6 =	9 × 6 =	7 × 6 =	0 × 6 =
4 × 6 =	1 × 6 =	6 × 6 =	3 × 6 =	8 × 6 =

Multiplying 7

9×7 6×7 5×7 2×7 7×7 8×7 4×7 3×7 0×7

4×7 1×7 6×7 2×7 5×7 3×7 8×7 9×7 7×7

1×7 3×7 8×7 0×7 9×7 4×7 6×7 7×7 2×7

7×7 4×7 6×7 8×7 1×7 5×7 3×7 0×7 9×7

3×7 1×7 7×7 6×7 0×7 9×7 2×7 4×7 8×7

5×7 0×7 2×7 9×7 4×7 7×7 1×7 3×7 6×7

6×7 9×7 3×7 7×7 0×7 8×7 2×7 5×7 1×7

8×7 2×7 5×7 9×7 1×7 4×7 0×7 7×7 3×7

0×7 5×7 9×7 7×7 4×7 2×7 8×7 6×7 1×7

Multiplying 8

8 × 8 =	4 × 8 =	3 × 8 =	0 × 8 =	1 × 8 =
9 × 8 =	5 × 8 =	6 × 8 =	2 × 8 =	7 × 8 =
4 × 8 =	6 × 8 =	7 × 8 =	5 × 8 =	2 × 8 =
1 × 8 =	3 × 8 =	8 × 8 =	0 × 8 =	9 × 8 =
2 × 8 =	8 × 8 =	6 × 8 =	1 × 8 =	3 × 8 =
0 × 8 =	5 × 8 =	9 × 8 =	7 × 8 =	4 × 8 =
5 × 8 =	3 × 8 =	0 × 8 =	9 × 8 =	2 × 8 =
7 × 8 =	4 × 8 =	6 × 8 =	8 × 8 =	1 × 8 =
9 × 8 =	2 × 8 =	8 × 8 =	4 × 8 =	5 × 8 =
3 × 8 =	1 × 8 =	7 × 8 =	6 × 8 =	0 × 8 =
7 × 8 =	3 × 8 =	1 × 8 =	8 × 8 =	6 × 8 =
5 × 8 =	0 × 8 =	2 × 8 =	9 × 8 =	4 × 8 =
8 × 8 =	2 × 8 =	5 × 8 =	3 × 8 =	1 × 8 =
6 × 8 =	9 × 8 =	3 × 8 =	7 × 8 =	0 × 8 =
4 × 8 =	0 × 8 =	7 × 8 =	3 × 8 =	6 × 8 =
8 × 8 =	2 × 8 =	5 × 8 =	9 × 8 =	1 × 8 =
5 × 8 =	8 × 8 =	9 × 8 =	7 × 8 =	0 × 8 =
4 × 8 =	1 × 8 =	6 × 8 =	3 × 8 =	6 × 8 =

Multiplying 9

9×9 | 6×9 | 5×9 | 2×9 | 7×9 | 8×9 | 4×9 | 3×9 | 0×9

4×9 | 1×9 | 6×9 | 2×9 | 5×9 | 3×9 | 8×9 | 9×9 | 7×9

1×9 | 3×9 | 8×9 | 0×9 | 9×9 | 4×9 | 6×9 | 7×9 | 2×9

7×9 | 4×9 | 6×9 | 8×9 | 1×9 | 5×9 | 3×9 | 0×9 | 9×9

3×9 | 1×9 | 7×9 | 6×9 | 0×9 | 9×9 | 2×9 | 4×9 | 8×9

5×9 | 0×9 | 2×9 | 9×9 | 4×9 | 7×9 | 1×9 | 3×9 | 6×9

6×9 | 9×9 | 3×9 | 7×9 | 0×9 | 8×9 | 2×9 | 5×9 | 1×9

8×9 | 2×9 | 5×9 | 9×9 | 1×9 | 4×9 | 0×9 | 7×9 | 3×9

0×9 | 5×9 | 9×9 | 7×9 | 4×9 | 2×9 | 8×9 | 6×9 | 1×9

Multiplying 10

$8 \times 10 =$	$4 \times 10 =$	$3 \times 10 =$	$0 \times 10 =$	$1 \times 10 =$
$9 \times 10 =$	$5 \times 10 =$	$6 \times 10 =$	$2 \times 10 =$	$7 \times 10 =$
$4 \times 10 =$	$6 \times 10 =$	$7 \times 10 =$	$5 \times 10 =$	$2 \times 10 =$
$1 \times 10 =$	$3 \times 10 =$	$8 \times 10 =$	$0 \times 10 =$	$9 \times 10 =$
$2 \times 10 =$	$8 \times 10 =$	$6 \times 10 =$	$1 \times 10 =$	$3 \times 10 =$
$0 \times 10 =$	$5 \times 10 =$	$9 \times 10 =$	$7 \times 10 =$	$4 \times 10 =$
$5 \times 10 =$	$3 \times 10 =$	$0 \times 10 =$	$9 \times 10 =$	$2 \times 10 =$
$7 \times 10 =$	$4 \times 10 =$	$6 \times 10 =$	$3 \times 10 =$	$1 \times 10 =$
$9 \times 10 =$	$2 \times 10 =$	$4 \times 10 =$	$8 \times 10 =$	$5 \times 10 =$
$3 \times 10 =$	$1 \times 10 =$	$7 \times 10 =$	$6 \times 10 =$	$0 \times 10 =$
$7 \times 10 =$	$3 \times 10 =$	$1 \times 10 =$	$8 \times 10 =$	$6 \times 10 =$
$5 \times 10 =$	$0 \times 10 =$	$2 \times 10 =$	$9 \times 10 =$	$4 \times 10 =$
$8 \times 10 =$	$2 \times 10 =$	$5 \times 10 =$	$3 \times 10 =$	$1 \times 10 =$
$6 \times 10 =$	$9 \times 10 =$	$3 \times 10 =$	$7 \times 10 =$	$0 \times 10 =$
$4 \times 10 =$	$0 \times 10 =$	$7 \times 10 =$	$3 \times 10 =$	$6 \times 10 =$
$8 \times 10 =$	$2 \times 10 =$	$5 \times 10 =$	$9 \times 10 =$	$1 \times 10 =$
$5 \times 10 =$	$8 \times 10 =$	$9 \times 10 =$	$7 \times 10 =$	$0 \times 10 =$
$4 \times 10 =$	$1 \times 10 =$	$6 \times 10 =$	$3 \times 10 =$	$6 \times 10 =$

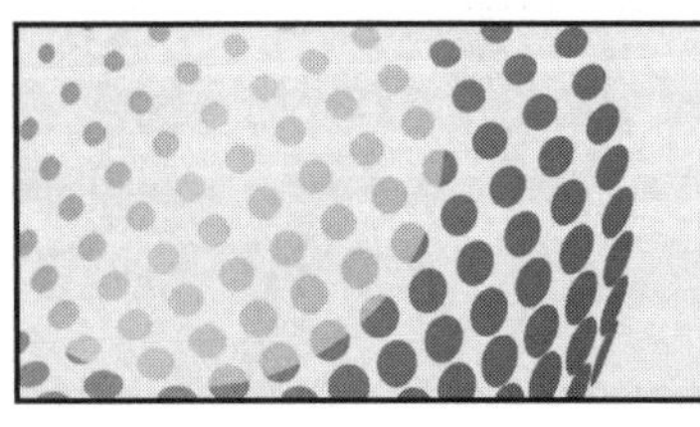

Section Diagnostic Test 6 – 10

7 × 6	5 × 7	5 × 8	1 × 9	0 ×10	5 × 6	6 × 7	7 × 8	4 × 9	3 ×10
9 × 6	1 × 7	8 × 8	9 × 9	2 ×10	2 × 6	8 × 7	1 × 8	8 × 9	0 10
8 × 6	3 × 7	6 × 8	7 × 9	5 ×10	0 × 6	6 × 7	0 × 8	9 × 9	7 ×10
5 × 6	2 × 7	4 × 8	5 × 9	6 ×10	6 × 6	4 × 7	3 × 8	7 × 9	5 ×10
1 × 6	4 × 7	2 × 8	3 × 9	4 ×10	4 × 6	2 × 7	5 × 8	6 × 9	2 ×10
0 × 6	0 × 7	0 × 8	2 × 9	8 ×10	8 × 6	0 × 7	6 × 8	7 × 9	4 ×10
4 × 6	5 × 7	1 × 8	4 × 9	9 ×10	6 × 6	7 × 7	9 × 8	4 × 9	8 ×10
3 × 6	7 × 7	3 × 8	6 × 9	7 ×10	2 × 6	3 × 7	8 × 8	6 × 9	6 ×10
2 × 6	9 × 7	5 × 8	8 × 9	4 ×10	9 × 6	6 × 7	4 × 8	9 × 9	3 ×10
5 × 6	6 × 7	7 × 8	0 × 9	1 ×10	0 × 6	5 × 7	2 × 8	5 × 9	9 ×10

Review Sheet

$7 \times 2 =$	$1 \times 6 =$	$4 \times 6 =$	$3 \times 5 =$	$3 \times 0 =$
$9 \times 6 =$	$9 \times 7 =$	$6 \times 6 =$	$2 \times 7 =$	$1 \times 4 =$
$7 \times 6 =$	$5 \times 3 =$	$5 \times 9 =$	$9 \times 3 =$	$4 \times 5 =$
$8 \times 8 =$	$4 \times 6 =$	$4 \times 3 =$	$7 \times 5 =$	$2 \times 9 =$
$7 \times 6 =$	$9 \times 0 =$	$0 \times 6 =$	$6 \times 8 =$	$7 \times 6 =$
$9 \times 9 =$	$8 \times 6 =$	$3 \times 7 =$	$2 \times 10 =$	$4 \times 8 =$
$4 \times 8 =$	$1 \times 5 =$	$5 \times 9 =$	$3 \times 8 =$	$2 \times 2 =$
$2 \times 4 =$	$4 \times 4 =$	$7 \times 8 =$	$8 \times 6 =$	$10 \times 5 =$
$7 \times 9 =$	$6 \times 2 =$	$3 \times 9 =$	$0 \times 0 =$	$2 \times 4 =$
$2 \times 2 =$	$6 \times 9 =$	$9 \times 9 =$	$10 \times 7 =$	$9 \times 8 =$
$3 \times 1 =$	$2 \times 8 =$	$8 \times 6 =$	$4 \times 8 =$	$5 \times 5 =$
$4 \times 9 =$	$3 \times 4 =$	$7 \times 3 =$	$4 \times 1 =$	$2 \times 9 =$
$2 \times 6 =$	$8 \times 9 =$	$9 \times 6 =$	$7 \times 4 =$	$4 \times 4 =$
$6 \times 3 =$	$8 \times 7 =$	$0 \times 8 =$	$3 \times 6 =$	$5 \times 8 =$
$7 \times 7 =$	$2 \times 5 =$	$4 \times 7 =$	$9 \times 4 =$	$9 \times 8 =$
$5 \times 7 =$	$7 \times 9 =$	$8 \times 5 =$	$6 \times 7 =$	$3 \times 3 =$
$5 \times 6 =$	$8 \times 8 =$	$2 \times 3 =$	$7 \times 7 =$	$4 \times 9 =$
$9 \times 6 =$	$9 \times 10 =$	$9 \times 8 =$	$6 \times 2 =$	$1 \times 4 =$

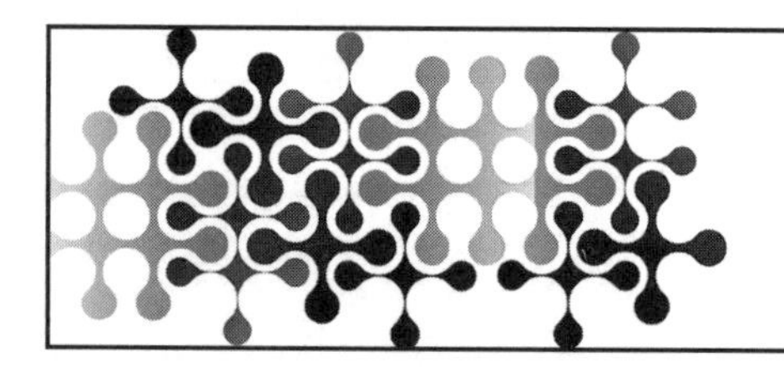

Review Sheet

7	___	6	3	___	6	7	___	3
× 5	× 6	× ___	× 8	× 7	× ___	× 2	× 3	× ___
___	42	48	___	49	30	___	18	21

___	5	4	___	3	10	___	5	4
× 8	× ___	× 9	× 8	× ___	× 9	× 2	× ___	× 7
32	40	___	72	30	___	6	25	___

10	6	___	2	6	___	10	9	___
× ___	× 4	× 9	× ___	× 3	× 4	× ___	× 7	× 4
60	___	54	8	___	20	0	___	12

3	___	6	5	___	7	9	___	9
× 3	×10	× ___	× 9	× 4	× ___	× 8	× 4	× ___
___	10	36	___	36	70	___	28	27

5 × 5 = ___   ___ × 6 = 42   10 × ___ = 40   ___ × 8 = 0

3 × ___ = 12   8 × 3 = ___   ___ × 2 = 16   6 × ___ = 48

___ × 7 = 63   9 × ___ = 90   7 × 7 = ___   ___ × 8 = 64

4 × 7 = ___   ___ × 4 = 32   9 × ___ = 81   6 × 2 = ___

2 × ___ = 10   8 × 5 = ___   ___ × 6 = 54   2 × ___ = 4

___ × 8 = 16   7 × ___ = 21   6 × 3 = ___   ___ × 2 = 18

2 × 7 = ___   ___ × 7 = 0   2 × ___ = 20   5 × 6 = ___

3 × ___ = 15   8 × 7 = ___   ___ × 1 = 1   8 × 9 = ___

___ × 9 = 45   6 × ___ = 36   10 × 7 = ___   ___ × 7 = 14

Final Assessment Test 0 – 10

0×9	5×8	2×7	6×5	9×6	5×2	3×1	3×0	7×0	8×10
9×3	1×9	8×8	9×7	3×5	7×6	8×2	2×1	5×0	8×0
3×4	8×3	2×9	0×8	3×7	7×5	3×6	7×2	5×1	6×0
9×2	5×4	7×3	3×9	3×8	8×7	9×5	5×6	6×2	4×1
2×2	4×2	4×4	6×3	4×9	7×8	4×7	4×5	8×6	9×2
8×8	3×3	8×2	2×4	5×3	5×9	4×8	5×7	8×5	4×6
9×1	1×1	7×7	3×2	9×4	4×3	6×9	4×8	6×7	5×5
0×10	6×1	5×5	6×6	0×2	7×4	3×3	7×9	6×8	7×7
2×10	6×10	7×1	9×9	0×0	6×2	8×4	2×3	8×9	9×8
5×10	7×10	9×10	8×1	4×4	5×5	7×2	6×4	1×3	9×9

Division

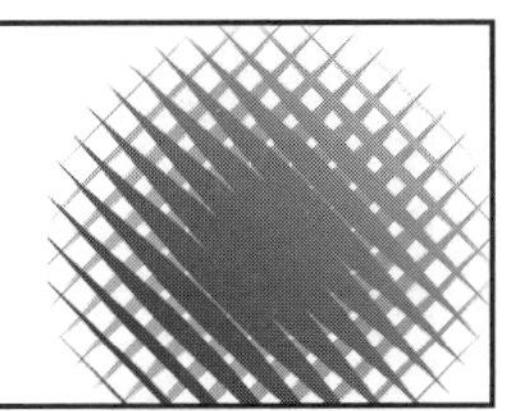

The **Beginning Assessment Test** has 100 problems, testing division facts 1–10. The Beginning Assessment Test helps to establish a simple knowledge base. Answers on page 88 show the diagonal arrangement of facts.

A periodic **Section Diagnostic Test** measures skill attainment to that point. You can make a judgement after each Section Diagnostic Test: Have skills within that section been mastered? Set a standard to move on to the next section. If that standard is not met, go back and focus on the problem areas(s) with Practice Sheets, or with similar materials that you prepare.

Some Practice Sheets provide problems in vertical form.

$2\overline{)16}$ $8\overline{)48}$ $5\overline{)25}$

The other Practice Sheets provide problems in horizontal form. Both forms are common.

$16 \div 2 =$ $48 \div 8 =$ $25 \div 5 =$

A periodic **Section Diagnostic Test** measures skill attainment to that point. You can make a judgement after each Section Diagnostic Test: Have skills within that section been mastered? Set a standard to move on to the next section. If that standard is not met, go back and focus on the problem areas(s) with Practice Sheets, or with similar materials that you prepare.

There are two **Review Sheets**. The first Review Sheet is straightforward. The second Review Sheet is different. Problems may require answers, as in $16 \div 2 =$ (quotient). Or, the Review Sheet may require a dividend or divisor to provide an answer (quotient).

$16 \div 2 =$ ___	___ $\div 8 = 6$	$25 \div$ ___ $= 5$
Dividend ÷ Divisor = Quotient	Dividend ÷ Divisor = Quotient	Dividend ÷ Divisor = Quotient

The **Final Assessment Test** provides 100 problems in a straightforward form. Again, the facts are arranged diagonally (see Answers, page 90). With 100 problems, each problem is equal to 1%.

Beginning Assessment Test

3)3	2)4	1)3	4)8	3)12	2)8	1)2	8)16	9)18	10)10
4)4	3)6	2)2	1)2	4)4	3)3	2)12	1)10	8)56	9)27
5)40	4)12	3)9	2)6	1)4	4)12	3)6	2)2	1)1	8)64
6)12	5)5	4)8	3)12	2)10	1)5	4)20	3)9	2)4	1)3
7)7	6)6	5)10	4)36	3)15	2)14	1)6	4)40	3)18	2)6
8)16	7)14	6)18	5)15	4)16	3)3	2)16	1)7	4)16	3)30
9)81	8)8	7)21	6)48	5)20	4)20	3)21	2)18	1)8	4)24
2)12	9)36	8)24	7)28	6)24	5)25	4)24	3)24	2)20	1)9
10)30	2)8	9)45	8)32	7)49	6)36	5)30	4)28	3)27	2)12
10)20	10)40	2)18	9)54	8)40	7)42	6)42	5)35	4)32	3)18

Dividing 1 & 2

$1\overline{)1}$	$2\overline{)2}$	$1\overline{)2}$	$2\overline{)4}$	$1\overline{)3}$	$2\overline{)6}$	$1\overline{)4}$	$2\overline{)8}$	$2\overline{)10}$
$1\overline{)6}$	$2\overline{)8}$	$1\overline{)7}$	$2\overline{)14}$	$1\overline{)8}$	$2\overline{)2}$	$1\overline{)9}$	$2\overline{)18}$	$1\overline{)10}$
$1\overline{)9}$	$1\overline{)1}$	$2\overline{)12}$	$2\overline{)16}$	$1\overline{)6}$	$1\overline{)7}$	$2\overline{)20}$	$1\overline{)10}$	$2\overline{)18}$
$2\overline{)20}$	$2\overline{)16}$	$1\overline{)3}$	$2\overline{)8}$	$1\overline{)5}$	$2\overline{)10}$	$1\overline{)1}$	$2\overline{)6}$	$1\overline{)7}$
$1\overline{)2}$	$2\overline{)12}$	$1\overline{)6}$	$1\overline{)1}$	$1\overline{)10}$	$2\overline{)10}$	$1\overline{)4}$	$2\overline{)4}$	$1\overline{)6}$
$2\overline{)20}$	$1\overline{)7}$	$2\overline{)18}$	$2\overline{)16}$	$1\overline{)9}$	$2\overline{)14}$	$1\overline{)10}$	$2\overline{)12}$	$1\overline{)10}$
$1\overline{)10}$	$1\overline{)1}$	$1\overline{)8}$	$1\overline{)2}$	$2\overline{)6}$	$1\overline{)3}$	$2\overline{)4}$	$1\overline{)2}$	$2\overline{)2}$
$2\overline{)4}$	$1\overline{)4}$	$2\overline{)6}$	$1\overline{)9}$	$2\overline{)8}$	$1\overline{)8}$	$2\overline{)10}$	$1\overline{)7}$	$2\overline{)12}$
$2\overline{)14}$	$1\overline{)5}$	$2\overline{)16}$	$1\overline{)4}$	$2\overline{)18}$	$1\overline{)3}$	$2\overline{)20}$	$1\overline{)2}$	$2\overline{)4}$

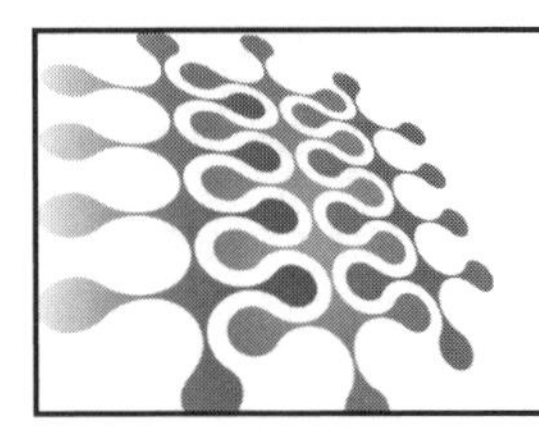

Dividing 3

3 ÷ 3 =	9 ÷ 3 =	12 ÷ 3 =	15 ÷ 3 =	27 ÷ 3 =
18 ÷ 3 =	6 ÷ 3 =	3 ÷ 3 =	18 ÷ 3 =	21 ÷ 3 =
15 ÷ 3 =	24 ÷ 3 =	12 ÷ 3 =	24 ÷ 3 =	9 ÷ 3 =
6 ÷ 3 =	12 ÷ 3 =	3 ÷ 3 =	15 ÷ 3 =	6 ÷ 3 =
27 ÷ 3 =	3 ÷ 3 =	21 ÷ 3 =	6 ÷ 3 =	21 ÷ 3 =
9 ÷ 3 =	21 ÷ 3 =	24 ÷ 3 =	18 ÷ 3 =	27 ÷ 3 =
18 ÷ 3 =	6 ÷ 3 =	15 ÷ 3 =	3 ÷ 3 =	12 ÷ 3 =
6 ÷ 3 =	18 ÷ 3 =	3 ÷ 3 =	24 ÷ 3 =	15 ÷ 3 =
12 ÷ 3 =	9 ÷ 3 =	30 ÷ 3 =	3 ÷ 3 =	9 ÷ 3 =
21 ÷ 3 =	15 ÷ 3 =	12 ÷ 3 =	30 ÷ 3 =	27 ÷ 3 =
6 ÷ 3 =	27 ÷ 3 =	24 ÷ 3 =	18 ÷ 3 =	24 ÷ 3 =
24 ÷ 3 =	18 ÷ 3 =	21 ÷ 3 =	9 ÷ 3 =	6 ÷ 3 =
3 ÷ 3 =	21 ÷ 3 =	18 ÷ 3 =	27 ÷ 3 =	12 ÷ 3 =
12 ÷ 3 =	3 ÷ 3 =	30 ÷ 3 =	15 ÷ 3 =	30 ÷ 3 =
15 ÷ 3 =	12 ÷ 3 =	9 ÷ 3 =	24 ÷ 3 =	9 ÷ 3 =
21 ÷ 3 =	18 ÷ 3 =	3 ÷ 3 =	27 ÷ 3 =	3 ÷ 3 =
30 ÷ 3 =	6 ÷ 3 =	30 ÷ 3 =	9 ÷ 3 =	18 ÷ 3 =
24 ÷ 3 =	27 ÷ 3 =	12 ÷ 3 =	30 ÷ 3 =	21 ÷ 3 =

Dividing 4

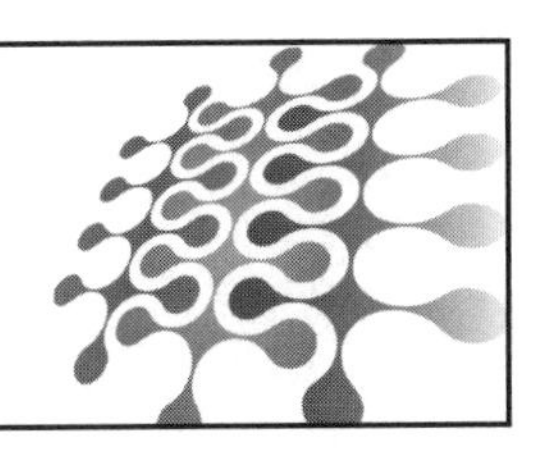

$4\overline{)4}$	$4\overline{)12}$	$4\overline{)24}$	$4\overline{)20}$	$4\overline{)40}$	$4\overline{)28}$	$4\overline{)16}$	$4\overline{)32}$	$4\overline{)8}$
$4\overline{)20}$	$4\overline{)12}$	$4\overline{)40}$	$4\overline{)36}$	$4\overline{)4}$	$4\overline{)24}$	$4\overline{)16}$	$4\overline{)4}$	$4\overline{)32}$
$4\overline{)36}$	$4\overline{)8}$	$4\overline{)12}$	$4\overline{)40}$	$4\overline{)16}$	$4\overline{)20}$	$4\overline{)12}$	$4\overline{)24}$	$4\overline{)8}$
$4\overline{)4}$	$4\overline{)40}$	$4\overline{)36}$	$4\overline{)8}$	$4\overline{)32}$	$4\overline{)16}$	$4\overline{)24}$	$4\overline{)28}$	$4\overline{)16}$
$4\overline{)8}$	$4\overline{)28}$	$4\overline{)20}$	$4\overline{)36}$	$4\overline{)4}$	$4\overline{)12}$	$4\overline{)40}$	$4\overline{)16}$	$4\overline{)24}$
$4\overline{)12}$	$4\overline{)32}$	$4\overline{)16}$	$4\overline{)4}$	$4\overline{)24}$	$4\overline{)8}$	$4\overline{)16}$	$4\overline{)40}$	$4\overline{)36}$
$4\overline{)32}$	$4\overline{)36}$	$4\overline{)40}$	$4\overline{)32}$	$4\overline{)36}$	$4\overline{)32}$	$4\overline{)40}$	$4\overline{)28}$	$4\overline{)24}$
$4\overline{)12}$	$4\overline{)4}$	$4\overline{)8}$	$4\overline{)28}$	$4\overline{)16}$	$4\overline{)32}$	$4\overline{)20}$	$4\overline{)8}$	$4\overline{)12}$
$4\overline{)28}$	$4\overline{)16}$	$4\overline{)8}$	$4\overline{)20}$	$4\overline{)4}$	$4\overline{)36}$	$4\overline{)24}$	$4\overline{)40}$	$4\overline{)16}$

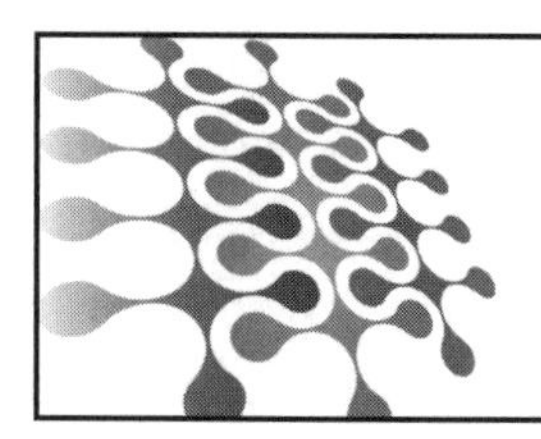

Dividing 5

10 ÷ 5 =	5 ÷ 5 =	45 ÷ 5 =	5 ÷ 5 =	35 ÷ 5 =
20 ÷ 5 =	25 ÷ 5 =	35 ÷ 5 =	40 ÷ 5 =	10 ÷ 5 =
15 ÷ 5 =	30 ÷ 5 =	10 ÷ 5 =	45 ÷ 5 =	50 ÷ 5 =
40 ÷ 5 =	50 ÷ 5 =	30 ÷ 5 =	15 ÷ 5 =	5 ÷ 5 =
5 ÷ 5 =	25 ÷ 5 =	20 ÷ 5 =	10 ÷ 5 =	15 ÷ 5 =
50 ÷ 5 =	40 ÷ 5 =	45 ÷ 5 =	5 ÷ 5 =	10 ÷ 5 =
10 ÷ 5 =	50 ÷ 5 =	15 ÷ 5 =	20 ÷ 5 =	25 ÷ 5 =
15 ÷ 5 =	35 ÷ 5 =	10 ÷ 5 =	25 ÷ 5 =	30 ÷ 5 =
25 ÷ 5 =	10 ÷ 5 =	5 ÷ 5 =	45 ÷ 5 =	5 ÷ 5 =
35 ÷ 5 =	5 ÷ 5 =	30 ÷ 5 =	10 ÷ 5 =	40 ÷ 5 =
20 ÷ 5 =	30 ÷ 5 =	35 ÷ 5 =	30 ÷ 5 =	20 ÷ 5 =
50 ÷ 5 =	10 ÷ 5 =	25 ÷ 5 =	20 ÷ 5 =	5 ÷ 5 =
45 ÷ 5 =	25 ÷ 5 =	15 ÷ 5 =	35 ÷ 5 =	40 ÷ 5 =
5 ÷ 5 =	50 ÷ 5 =	35 ÷ 5 =	30 ÷ 5 =	10 ÷ 5 =
35 ÷ 5 =	40 ÷ 5 =	50 ÷ 5 =	15 ÷ 5 =	5 ÷ 5 =
40 ÷ 5 =	15 ÷ 5 =	25 ÷ 5 =	50 ÷ 5 =	25 ÷ 5 =
10 ÷ 5 =	45 ÷ 5 =	40 ÷ 5 =	10 ÷ 5 =	45 ÷ 5 =
15 ÷ 5 =	20 ÷ 5 =	25 ÷ 5 =	20 ÷ 5 =	15 ÷ 5 =

Section Diagnostic Test 1 – 5

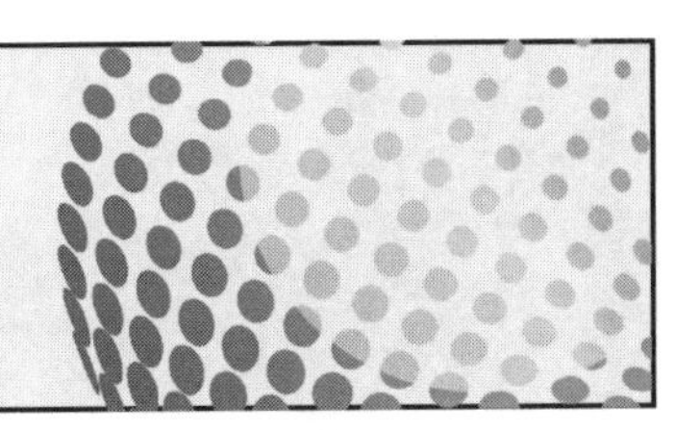

$10 \div 5 =$	$5 \div 5 =$	$45 \div 5 =$	$5 \div 5 =$	$35 \div 5 =$
$20 \div 4 =$	$24 \div 4 =$	$36 \div 4 =$	$16 \div 4 =$	$40 \div 4 =$
$15 \div 3 =$	$27 \div 3 =$	$6 \div 3 =$	$3 \div 3 =$	$30 \div 3 =$
$14 \div 2 =$	$4 \div 2 =$	$20 \div 2 =$	$10 \div 2 =$	$8 \div 2 =$
$1 \div 1 =$	$4 \div 1 =$	$7 \div 1 =$	$2 \div 1 =$	$3 \div 1 =$
$50 \div 5 =$	$40 \div 5 =$	$45 \div 5 =$	$5 \div 5 =$	$10 \div 5 =$
$4 \div 4 =$	$36 \div 4 =$	$12 \div 4 =$	$20 \div 4 =$	$8 \div 4 =$
$6 \div 3 =$	$15 \div 3 =$	$21 \div 3 =$	$27 \div 3 =$	$3 \div 3 =$
$4 \div 2 =$	$8 \div 2 =$	$16 \div 2 =$	$20 \div 2 =$	$18 \div 2 =$
$25 \div 5 =$	$5 \div 5 =$	$30 \div 5 =$	$35 \div 5 =$	$40 \div 5 =$
$20 \div 4 =$	$28 \div 4 =$	$20 \div 4 =$	$16 \div 4 =$	$32 \div 4 =$
$30 \div 3 =$	$27 \div 3 =$	$12 \div 3 =$	$18 \div 3 =$	$24 \div 3 =$
$10 \div 2 =$	$12 \div 2 =$	$14 \div 2 =$	$8 \div 2 =$	$6 \div 2 =$
$8 \div 1 =$	$6 \div 1 =$	$10 \div 1 =$	$4 \div 1 =$	$7 \div 1 =$
$35 \div 5 =$	$40 \div 5 =$	$50 \div 5 =$	$15 \div 5 =$	$5 \div 5 =$
$40 \div 4 =$	$8 \div 4 =$	$12 \div 4 =$	$24 \div 4 =$	$16 \div 4 =$
$15 \div 3 =$	$21 \div 3 =$	$18 \div 3 =$	$3 \div 3 =$	$6 \div 3 =$
$6 \div 2 =$	$20 \div 2 =$	$10 \div 2 =$	$14 \div 2 =$	$2 \div 2 =$

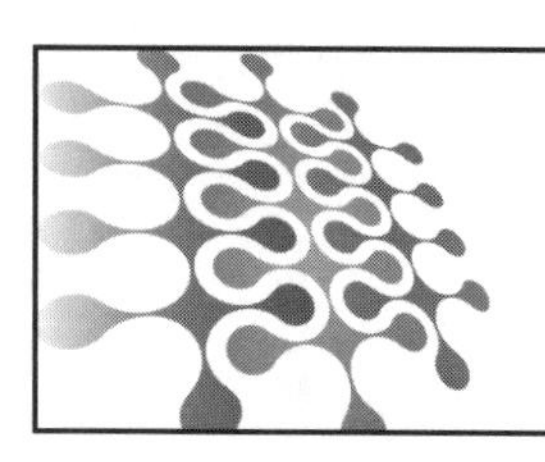

Dividing 6

$6\overline{)6}$ $6\overline{)18}$ $6\overline{)12}$ $6\overline{)30}$ $6\overline{)54}$ $6\overline{)24}$ $6\overline{)60}$ $6\overline{)36}$ $6\overline{)42}$

$6\overline{)18}$ $6\overline{)36}$ $6\overline{)48}$ $6\overline{)6}$ $6\overline{)30}$ $6\overline{)54}$ $6\overline{)42}$ $6\overline{)12}$ $6\overline{)60}$

$6\overline{)6}$ $6\overline{)30}$ $6\overline{)12}$ $6\overline{)54}$ $6\overline{)6}$ $6\overline{)48}$ $6\overline{)24}$ $6\overline{)42}$ $6\overline{)18}$

$6\overline{)30}$ $6\overline{)12}$ $6\overline{)54}$ $6\overline{)6}$ $6\overline{)48}$ $6\overline{)24}$ $6\overline{)60}$ $6\overline{)18}$ $6\overline{)36}$

$6\overline{)24}$ $6\overline{)6}$ $6\overline{)42}$ $6\overline{)12}$ $6\overline{)60}$ $6\overline{)48}$ $6\overline{)18}$ $6\overline{)6}$ $6\overline{)30}$

$6\overline{)60}$ $6\overline{)48}$ $6\overline{)54}$ $6\overline{)18}$ $6\overline{)36}$ $6\overline{)42}$ $6\overline{)30}$ $6\overline{)24}$ $6\overline{)12}$

$6\overline{)48}$ $6\overline{)24}$ $6\overline{)6}$ $6\overline{)54}$ $6\overline{)12}$ $6\overline{)60}$ $6\overline{)36}$ $6\overline{)42}$ $6\overline{)18}$

$6\overline{)36}$ $6\overline{)42}$ $6\overline{)18}$ $6\overline{)24}$ $6\overline{)60}$ $6\overline{)54}$ $6\overline{)48}$ $6\overline{)30}$ $6\overline{)6}$

$6\overline{)12}$ $6\overline{)54}$ $6\overline{)30}$ $6\overline{)36}$ $6\overline{)42}$ $6\overline{)18}$ $6\overline{)6}$ $6\overline{)48}$ $6\overline{)24}$

Dividing 7

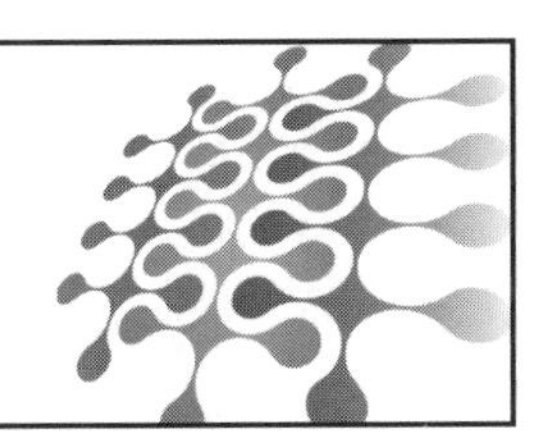

$7 \div 7 =$	$35 \div 7 =$	$70 \div 7 =$	$42 \div 7 =$	$56 \div 7 =$
$21 \div 7 =$	$63 \div 7 =$	$35 \div 7 =$	$49 \div 7 =$	$28 \div 7 =$
$28 \div 7 =$	$21 \div 7 =$	$7 \div 7 =$	$70 \div 7 =$	$42 \div 7 =$
$49 \div 7 =$	$35 \div 7 =$	$63 \div 7 =$	$56 \div 7 =$	$14 \div 7 =$
$14 \div 7 =$	$42 \div 7 =$	$70 \div 7 =$	$28 \div 7 =$	$63 \div 7 =$
$7 \div 7 =$	$14 \div 7 =$	$35 \div 7 =$	$63 \div 7 =$	$49 \div 7 =$
$56 \div 7 =$	$49 \div 7 =$	$42 \div 7 =$	$35 \div 7 =$	$14 \div 7 =$
$35 \div 7 =$	$28 \div 7 =$	$21 \div 7 =$	$70 \div 7 =$	$7 \div 7 =$
$42 \div 7 =$	$56 \div 7 =$	$14 \div 7 =$	$21 \div 7 =$	$42 \div 7 =$
$63 \div 7 =$	$7 \div 7 =$	$28 \div 7 =$	$7 \div 7 =$	$21 \div 7 =$
$14 \div 7 =$	$21 \div 7 =$	$63 \div 7 =$	$49 \div 7 =$	$28 \div 7 =$
$42 \div 7 =$	$63 \div 7 =$	$70 \div 7 =$	$14 \div 7 =$	$56 \div 7 =$
$49 \div 7 =$	$7 \div 7 =$	$35 \div 7 =$	$70 \div 7 =$	$35 \div 7 =$
$21 \div 7 =$	$49 \div 7 =$	$42 \div 7 =$	$14 \div 7 =$	$28 \div 7 =$
$35 \div 7 =$	$70 \div 7 =$	$7 \div 7 =$	$56 \div 7 =$	$21 \div 7 =$
$42 \div 7 =$	$7 \div 7 =$	$21 \div 7 =$	$28 \div 7 =$	$35 \div 7 =$
$28 \div 7 =$	$42 \div 7 =$	$7 \div 7 =$	$14 \div 7 =$	$49 \div 7 =$
$7 \div 7 =$	$14 \div 7 =$	$49 \div 7 =$	$35 \div 7 =$	$21 \div 7 =$

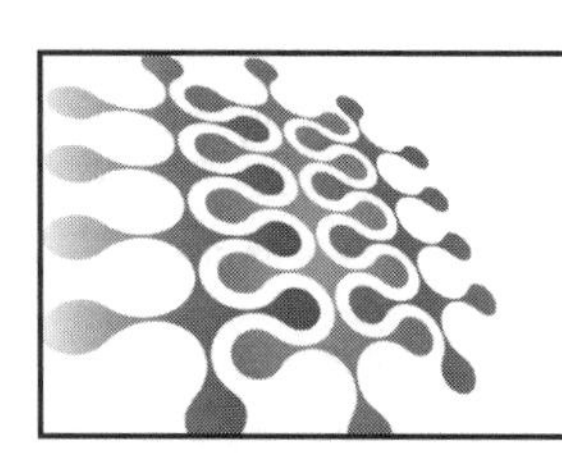

Dividing 8

$8\overline{)8}$ $8\overline{)24}$ $8\overline{)16}$ $8\overline{)32}$ $8\overline{)48}$ $8\overline{)72}$ $8\overline{)80}$ $8\overline{)56}$ $8\overline{)64}$

$8\overline{)24}$ $8\overline{)56}$ $8\overline{)40}$ $8\overline{)16}$ $8\overline{)72}$ $8\overline{)48}$ $8\overline{)8}$ $8\overline{)48}$ $8\overline{)32}$

$8\overline{)32}$ $8\overline{)64}$ $8\overline{)8}$ $8\overline{)48}$ $8\overline{)40}$ $8\overline{)16}$ $8\overline{)56}$ $8\overline{)80}$ $8\overline{)24}$

$8\overline{)40}$ $8\overline{)80}$ $8\overline{)64}$ $8\overline{)32}$ $8\overline{)8}$ $8\overline{)24}$ $8\overline{)48}$ $8\overline{)16}$ $8\overline{)56}$

$8\overline{)16}$ $8\overline{)24}$ $8\overline{)80}$ $8\overline{)72}$ $8\overline{)40}$ $8\overline{)64}$ $8\overline{)56}$ $8\overline{)8}$ $8\overline{)40}$

$8\overline{)80}$ $8\overline{)32}$ $8\overline{)8}$ $8\overline{)64}$ $8\overline{)72}$ $8\overline{)48}$ $8\overline{)40}$ $8\overline{)16}$ $8\overline{)24}$

$8\overline{)64}$ $8\overline{)40}$ $8\overline{)24}$ $8\overline{)48}$ $8\overline{)40}$ $8\overline{)72}$ $8\overline{)8}$ $8\overline{)80}$ $8\overline{)16}$

$8\overline{)8}$ $8\overline{)72}$ $8\overline{)32}$ $8\overline{)8}$ $8\overline{)56}$ $8\overline{)16}$ $8\overline{)24}$ $8\overline{)72}$ $8\overline{)24}$

$8\overline{)72}$ $8\overline{)48}$ $8\overline{)24}$ $8\overline{)80}$ $8\overline{)16}$ $8\overline{)8}$ $8\overline{)32}$ $8\overline{)24}$ $8\overline{)40}$

Dividing 9

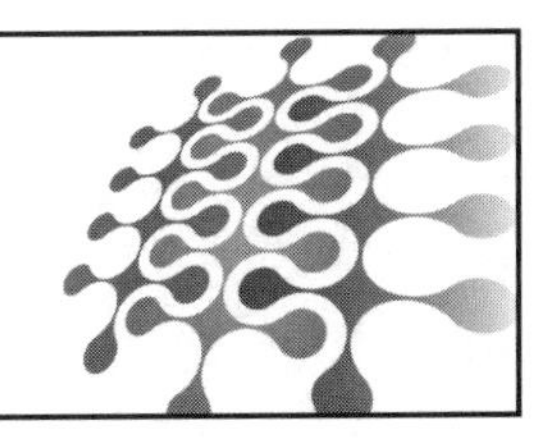

$9 \div 9 =$	$36 \div 9 =$	$27 \div 9 =$	$81 \div 9 =$	$72 \div 9 =$
$18 \div 9 =$	$63 \div 9 =$	$9 \div 9 =$	$72 \div 9 =$	$81 \div 9 =$
$45 \div 9 =$	$81 \div 9 =$	$54 \div 9 =$	$27 \div 9 =$	$90 \div 9 =$
$27 \div 9 =$	$36 \div 9 =$	$90 \div 9 =$	$9 \div 9 =$	$18 \div 9 =$
$63 \div 9 =$	$18 \div 9 =$	$72 \div 9 =$	$36 \div 9 =$	$45 \div 9 =$
$36 \div 9 =$	$27 \div 9 =$	$81 \div 9 =$	$90 \div 9 =$	$9 \div 9 =$
$45 \div 9 =$	$9 \div 9 =$	$63 \div 9 =$	$72 \div 9 =$	$54 \div 9 =$
$90 \div 9 =$	$45 \div 9 =$	$36 \div 9 =$	$18 \div 9 =$	$27 \div 9 =$
$72 \div 9 =$	$54 \div 9 =$	$18 \div 9 =$	$45 \div 9 =$	$72 \div 9 =$
$18 \div 9 =$	$45 \div 9 =$	$27 \div 9 =$	$63 \div 9 =$	$18 \div 9 =$
$90 \div 9 =$	$9 \div 9 =$	$54 \div 9 =$	$36 \div 9 =$	$9 \div 9 =$
$81 \div 9 =$	$90 \div 9 =$	$9 \div 9 =$	$63 \div 9 =$	$72 \div 9 =$
$45 \div 9 =$	$72 \div 9 =$	$54 \div 9 =$	$18 \div 9 =$	$36 \div 9 =$
$18 \div 9 =$	$9 \div 9 =$	$90 \div 9 =$	$81 \div 9 =$	$45 \div 9 =$
$81 \div 9 =$	$36 \div 9 =$	$63 \div 9 =$	$27 \div 9 =$	$72 \div 9 =$
$9 \div 9 =$	$72 \div 9 =$	$81 \div 9 =$	$18 \div 9 =$	$63 \div 9 =$
$27 \div 9 =$	$45 \div 9 =$	$18 \div 9 =$	$90 \div 9 =$	$36 \div 9 =$
$36 \div 9 =$	$90 \div 9 =$	$27 \div 9 =$	$81 \div 9 =$	$45 \div 9 =$

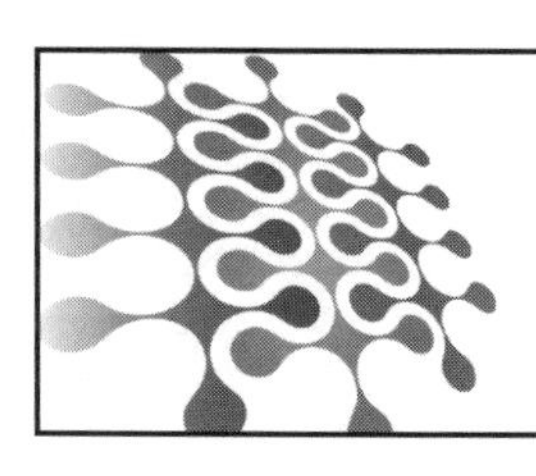

Dividing 10

$10\overline{)10}$ $10\overline{)90}$ $10\overline{)50}$ $10\overline{)80}$ $10\overline{)30}$ $10\overline{)70}$ $10\overline{)100}$ $10\overline{)20}$ $10\overline{)40}$

$10\overline{)50}$ $10\overline{)20}$ $10\overline{)60}$ $10\overline{)10}$ $10\overline{)40}$ $10\overline{)90}$ $10\overline{)30}$ $10\overline{)100}$ $10\overline{)70}$

$10\overline{)20}$ $10\overline{)100}$ $10\overline{)70}$ $10\overline{)30}$ $10\overline{)80}$ $10\overline{)50}$ $10\overline{)90}$ $10\overline{)10}$ $10\overline{)40}$

$10\overline{)80}$ $10\overline{)30}$ $10\overline{)20}$ $10\overline{)70}$ $10\overline{)90}$ $10\overline{)60}$ $10\overline{)40}$ $10\overline{)100}$ $10\overline{)10}$

$10\overline{)60}$ $10\overline{)20}$ $10\overline{)40}$ $10\overline{)90}$ $10\overline{)100}$ $10\overline{)80}$ $10\overline{)10}$ $10\overline{)50}$ $10\overline{)30}$

$10\overline{)100}$ $10\overline{)40}$ $10\overline{)90}$ $10\overline{)60}$ $10\overline{)70}$ $10\overline{)20}$ $10\overline{)30}$ $10\overline{)80}$ $10\overline{)50}$

$10\overline{)50}$ $10\overline{)10}$ $10\overline{)80}$ $10\overline{)100}$ $10\overline{)20}$ $10\overline{)90}$ $10\overline{)40}$ $10\overline{)60}$ $10\overline{)90}$

$10\overline{)40}$ $10\overline{)30}$ $10\overline{)20}$ $10\overline{)10}$ $10\overline{)50}$ $10\overline{)100}$ $10\overline{)60}$ $10\overline{)90}$ $10\overline{)70}$

$10\overline{)90}$ $10\overline{)80}$ $10\overline{)100}$ $10\overline{)60}$ $10\overline{)70}$ $10\overline{)10}$ $10\overline{)20}$ $10\overline{)30}$ $10\overline{)40}$

Section Diagnostic Test 6 – 10

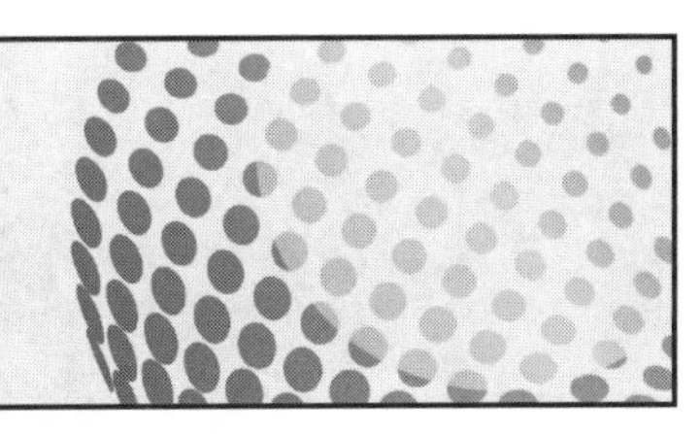

42 ÷ 6 =	54 ÷ 6 =	48 ÷ 6 =	30 ÷ 6 =	36 ÷ 6 =
7 ÷ 7 =	21 ÷ 7 =	14 ÷ 7 =	28 ÷ 7 =	70 ÷ 7 =
24 ÷ 8 =	48 ÷ 8 =	64 ÷ 8 =	8 ÷ 8 =	32 ÷ 8 =
81 ÷ 9 =	45 ÷ 9 =	9 ÷ 9 =	54 ÷ 9 =	27 ÷ 9 =
6 ÷ 6 =	60 ÷ 6 =	24 ÷ 6 =	18 ÷ 6 =	12 ÷ 6 =
35 ÷ 7 =	49 ÷ 7 =	63 ÷ 7 =	56 ÷ 7 =	42 ÷ 7 =
80 ÷ 8 =	48 ÷ 8 =	16 ÷ 8 =	32 ÷ 8 =	64 ÷ 8 =
27 ÷ 9 =	36 ÷ 9 =	18 ÷ 9 =	45 ÷ 9 =	63 ÷ 9 =
90 ÷ 10 =	20 ÷ 10 =	50 ÷ 10 =	100 ÷ 10 =	70 ÷ 10 =
72 ÷ 9 =	54 ÷ 9 =	27 ÷ 9 =	90 ÷ 9 =	72 ÷ 9 =
32 ÷ 8 =	40 ÷ 8 =	32 ÷ 8 =	16 ÷ 8 =	8 ÷ 8 =
63 ÷ 7 =	49 ÷ 7 =	35 ÷ 7 =	21 ÷ 7 =	7 ÷ 7 =
18 ÷ 6 =	6 ÷ 6 =	54 ÷ 6 =	30 ÷ 6 =	42 ÷ 6 =
10 ÷ 5 =	5 ÷ 5 =	45 ÷ 5 =	5 ÷ 5 =	35 ÷ 5 =
12 ÷ 6 =	60 ÷ 6 =	30 ÷ 6 =	24 ÷ 6 =	48 ÷ 6 =
14 ÷ 7 =	21 ÷ 7 =	35 ÷ 7 =	7 ÷ 7 =	70 ÷ 7 =
64 ÷ 8 =	24 ÷ 8 =	40 ÷ 8 =	16 ÷ 8 =	8 ÷ 8 =
81 ÷ 9 =	72 ÷ 9 =	45 ÷ 9 =	18 ÷ 9 =	27 ÷ 9 =

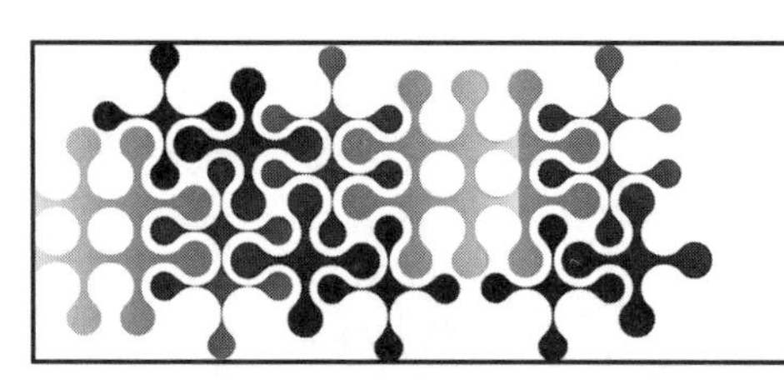

Review Sheet

$3\overline{)3}$ $2\overline{)16}$ $8\overline{)48}$ $2\overline{)16}$ $5\overline{)25}$ $4\overline{)36}$ $3\overline{)12}$ $7\overline{)21}$ $4\overline{)4}$

$2\overline{)18}$ $8\overline{)72}$ $2\overline{)12}$ $9\overline{)54}$ $7\overline{)28}$ $4\overline{)16}$ $6\overline{)18}$ $8\overline{)56}$ $8\overline{)80}$

$3\overline{)18}$ $5\overline{)40}$ $7\overline{)49}$ $2\overline{)10}$ $4\overline{)28}$ $9\overline{)36}$ $9\overline{)72}$ $5\overline{)35}$ $7\overline{)63}$

$8\overline{)40}$ $6\overline{)42}$ $3\overline{)9}$ $5\overline{)30}$ $9\overline{)72}$ $2\overline{)6}$ $7\overline{)49}$ $4\overline{)36}$ $9\overline{)54}$

$5\overline{)50}$ $9\overline{)72}$ $6\overline{)12}$ $1\overline{)1}$ $3\overline{)9}$ $8\overline{)24}$ $6\overline{)24}$ $5\overline{)10}$ $7\overline{)63}$

$7\overline{)42}$ $6\overline{)36}$ $3\overline{)6}$ $7\overline{)70}$ $8\overline{)16}$ $7\overline{)14}$ $1\overline{)6}$ $4\overline{)24}$ $3\overline{)15}$

$3\overline{)30}$ $9\overline{)54}$ $9\overline{)63}$ $6\overline{)36}$ $2\overline{)14}$ $1\overline{)4}$ $7\overline{)42}$ $5\overline{)15}$ $5\overline{)45}$

$9\overline{)27}$ $4\overline{)20}$ $8\overline{)64}$ $4\overline{)24}$ $4\overline{)12}$ $7\overline{)35}$ $2\overline{)18}$ $9\overline{)90}$ $6\overline{)48}$

$9\overline{)81}$ $8\overline{)48}$ $3\overline{)21}$ $2\overline{)20}$ $4\overline{)32}$ $5\overline{)45}$ $3\overline{)24}$ $8\overline{)32}$ $7\overline{)56}$

Review Sheet

1		6	3		10	4		7
___) 5	6) 30	7) ___	___) 24	3) 27	10) ___	___) 32	5) 35	7) ___

	8	9		7	4		10	6
5) 35	9) ___	___) 36	2) 12	8) ___	___) 28	8) 40	2) ___	___) 36

8	1		8	7		1	9	
4) ___	___) 7	6) 18	5) ___	___) 21	9) 81	3) ___	___) 63	8) 56

3		6	4		6	9		4
___) 24	7) 49	9) ___	___) 12	2) 18	8) ___	___) 36	9) 45	5) ___

30 ÷ 5 = ___	___ ÷ 6 = 7	40 ÷ ___ = 4	___ ÷ 8 = 3
63 ÷ ___ = 7	32 ÷ 4 = ___	___ ÷ 2 = 4	18 ÷ ___ = 6
___ ÷ 5 = 9	90 ÷ ___ = 9	49 ÷ 7 = ___	___ ÷ 8 = 8
36 ÷ 9 = ___	___ ÷ 4 = 3	72 ÷ ___ = 8	___ ÷ 6 = 5
10 ÷ ___ = 2	50 ÷ 5 = ___	___ ÷ 6 = 9	16 ÷ ___ = 4
___ ÷ 8 = 2	21 ÷ ___ = 7	27 ÷ 3 = ___	___ ÷ 2 = 9
100 ÷ 10 = ___	___ ÷ 5 = 4	32 ÷ ___ = 4	___ ÷ 3 = 9
12 ÷ ___ = 2	56 ÷ 7 = ___	___ ÷ 1 = 1	30 ÷ ___ = 6
___ ÷ 3 = 7	14 ÷ ___ = 7	24 ÷ 4 = ___	___ ÷ 4 = 8

Final Assessment Test 0 – 10

$9\overline{)18}$ $8\overline{)16}$ $7\overline{)7}$ $6\overline{)60}$ $5\overline{)20}$ $4\overline{)16}$ $5\overline{)5}$ $6\overline{)18}$ $7\overline{)28}$ $10\overline{)100}$

$3\overline{)12}$ $9\overline{)54}$ $8\overline{)48}$ $7\overline{)35}$ $6\overline{)12}$ $5\overline{)30}$ $4\overline{)28}$ $5\overline{)10}$ $6\overline{)36}$ $7\overline{)21}$

$4\overline{)8}$ $3\overline{)3}$ $9\overline{)72}$ $8\overline{)72}$ $7\overline{)14}$ $6\overline{)54}$ $5\overline{)45}$ $4\overline{)20}$ $5\overline{)50}$ $6\overline{)48}$

$2\overline{)6}$ $4\overline{)12}$ $3\overline{)9}$ $9\overline{)27}$ $8\overline{)64}$ $7\overline{)63}$ $6\overline{)30}$ $5\overline{)15}$ $4\overline{)36}$ $5\overline{)40}$

$1\overline{)3}$ $2\overline{)4}$ $4\overline{)40}$ $3\overline{)30}$ $9\overline{)81}$ $8\overline{)32}$ $7\overline{)70}$ $6\overline{)48}$ $5\overline{)35}$ $4\overline{)24}$

$3\overline{)27}$ $1\overline{)4}$ $2\overline{)2}$ $4\overline{)32}$ $3\overline{)15}$ $9\overline{)36}$ $8\overline{)8}$ $7\overline{)42}$ $6\overline{)24}$ $5\overline{)10}$

$8\overline{)8}$ $3\overline{)18}$ $1\overline{)7}$ $2\overline{)8}$ $4\overline{)12}$ $3\overline{)27}$ $9\overline{)18}$ $8\overline{)40}$ $7\overline{)56}$ $6\overline{)42}$

$2\overline{)18}$ $8\overline{)56}$ $3\overline{)30}$ $1\overline{)5}$ $2\overline{)10}$ $4\overline{)36}$ $3\overline{)18}$ $9\overline{)9}$ $8\overline{)80}$ $7\overline{)49}$

$10\overline{)60}$ $2\overline{)20}$ $8\overline{)72}$ $3\overline{)21}$ $1\overline{)6}$ $2\overline{)16}$ $4\overline{)24}$ $3\overline{)21}$ $9\overline{)18}$ $8\overline{)24}$

$10\overline{)80}$ $10\overline{)30}$ $2\overline{)14}$ $8\overline{)48}$ $3\overline{)6}$ $1\overline{)8}$ $2\overline{)12}$ $4\overline{)28}$ $3\overline{)9}$ $9\overline{)90}$

Answers
Addition

Beginning Assessment Test, page 8.

0 + 9 = 9	9 + 8 = 17	6 + 7 = 13	5 + 4 = 9	1 + 3 = 4	0 + 2 = 2	2 + 1 = 3	9 + 0 = 9	3 + 0 = 3	5 + 10 = 15	10s
9 + 6 = 15	1 + 9 = 10	7 + 8 = 15	3 + 7 = 10	9 + 4 = 13	4 + 3 = 7	5 + 2 = 7	4 + 1 = 5	8 + 0 = 8	1 + 0 = 1	0s
4 + 5 = 9	8 + 6 = 14	2 + 9 = 11	6 + 8 = 14	1 + 7 = 8	4 + 4 = 8	5 + 3 = 8	3 + 2 = 5	8 + 1 = 9	7 + 0 = 7	0s
10 + 4 = 14	2 + 5 = 7	7 + 6 = 13	3 + 9 = 12	5 + 8 = 13	8 + 7 = 15	7 + 4 = 11	6 + 3 = 9	6 + 2 = 8	3 + 1 = 4	1s
0 + 0 = 0	2 + 4 = 6	9 + 5 = 14	10 + 6 = 16	4 + 9 = 13	4 + 8 = 12	2 + 7 = 9	8 + 4 = 12	2 + 3 = 5	7 + 2 = 9	2s
8 + 8 = 16	3 + 3 = 6	8 + 4 = 12	6 + 5 = 11	4 + 6 = 10	5 + 9 = 14	3 + 8 = 11	9 + 7 = 16	5 + 4 = 9	8 + 3 = 11	3s
7 + 3 = 10	1 + 1 = 2	6 + 6 = 12	7 + 4 = 11	3 + 5 = 8	3 + 6 = 9	6 + 9 = 15	2 + 8 = 10	10 + 7 = 17	2 + 4 = 6	4s
8 + 2 = 10	5 + 3 = 8	2 + 2 = 4	9 + 9 = 18	3 + 4 = 7	8 + 5 = 13	2 + 6 = 8	7 + 9 = 16	1 + 8 = 9	5 + 7 = 12	7s
0 + 10 = 10	9 + 2 = 11	10 + 3 = 13	5 + 5 = 10	10 + 10 = 20	1 + 4 = 5	1 + 5 = 6	1 + 6 = 7	8 + 9 = 17	0 + 8 = 8	8s
9 + 10 = 19	8 + 10 = 18	10 + 2 = 12	0 + 3 = 3	4 + 4 = 8	7 + 7 = 14	0 + 4 = 4	7 + 5 = 12	0 + 6 = 6	9 + 9 = 18	9s
10s	10s	2s	3s	DOUBLES	4s	5s	6s			

The **Beginning Assessment Test** has facts arranged diagonally. This diagonal arrangement quickly identifies facts that are firm and facts that need attention.

Begin *Practice Sheets* at the level where several errors occur in fact diagonals. That may be with 2s for younger learners and 4s with older learners.

Practice Sheet 0s & 1s, page 9.

8	5	4	4	11	2	6	8	11
2	8	3	5	6	8	0	7	7
1	7	10	8	7	2	0	2	8
11	4	2	0	11	3	4	10	4
4	6	9	3	8	4	1	9	10
5	10	6	1	6	5	0	2	1
2	3	5	10	3	2	3	5	10
1	9	7	9	9	1	9	1	6
10	5	9	3	4	7	10	8	2

Practice Sheet 2s, page 10.

4	10	6	8	5
3	2	12	9	11
11	8	7	4	9
10	6	5	2	3
3	5	10	8	11
12	8	9	4	12
2	7	11	12	6
4	10	8	3	5
9	6	9	10	3
7	5	2	11	4
5	12	9	8	2
11	4	6	10	7
7	2	4	11	6
9	3	5	8	10
8	11	10	9	12
10	4	7	3	6
5	8	12	11	3
6	2	9	5	8

Practice Sheet 3s, page 11.

9	8	5	10	11	7	6	3	4
6	11	3	12	13	9	10	5	13
8	12	13	7	5	11	9	4	6
7	9	11	4	8	6	3	12	5
13	10	9	3	12	13	7	11	8
3	5	12	7	10	4	6	9	11
12	6	10	13	11	5	8	4	7
5	13	12	4	7	3	10	6	9
4	9	5	8	6	11	12	13	3

Practice Sheet 4s, page 12.

8	7	14	6	9
10	11	12	7	13
7	4	5	14	5
5	13	9	11	8
4	8	7	5	12
9	10	4	7	11
14	13	14	10	8
13	6	12	6	10
5	9	14	8	4
4	7	12	5	11
7	5	8	9	13
8	10	6	4	14
14	12	13	14	7
11	6	4	9	4
6	8	10	14	6
9	11	7	11	10
10	13	5	13	6
13	4	7	10	8

Practice Sheet 5s, page 13.

13	9	11	8	6	5	15	12	14
11	10	7	12	13	9	8	5	6
8	13	5	14	15	11	12	7	15
10	14	15	9	7	13	11	6	8
9	11	13	6	10	8	5	14	7
15	12	11	5	14	15	9	13	10
5	7	14	9	12	6	8	11	14
14	8	12	15	13	7	10	6	9
7	15	14	6	9	5	12	8	11

Answers
Addition

Section Diagnostic Test 0–5, page 14.

5s	4s	3s	2s	1s	0s	2s	3s	4s	5s
0 + 5 = 5	5 + 4 = 9	2 + 3 = 5	2 + 2 = 4	4 + 1 = 5	1 + 0 = 1	5 + 2 = 7	3 + 3 = 6	4 + 4 = 8	5 + 5 = 10
2 + 5 = 7	0 + 4 = 4	5 + 3 = 8	3 + 2 = 5	3 + 1 = 4	4 + 0 = 4	1 + 2 = 3	5 + 3 = 8	3 + 4 = 7	4 + 5 = 9
4 + 5 = 9	2 + 4 = 6	0 + 3 = 3	5 + 2 = 7	2 + 1 = 3	3 + 0 = 3	4 + 2 = 6	1 + 3 = 4	5 + 4 = 9	3 + 5 = 8
3 + 5 = 8	4 + 4 = 8	2 + 3 = 5	0 + 2 = 2	5 + 1 = 6	2 + 0 = 2	3 + 2 = 5	4 + 3 = 7	1 + 4 = 5	5 + 5 = 10
5 + 5 = 10	3 + 4 = 7	4 + 3 = 7	2 + 2 = 4	0 + 1 = 1	5 + 0 = 5	2 + 2 = 4	3 + 3 = 6	4 + 4 = 8	1 + 5 = 6
1 + 5 = 6	5 + 4 = 9	3 + 3 = 6	4 + 2 = 6	2 + 1 = 3	0 + 0 = 0	5 + 2 = 7	2 + 3 = 5	3 + 4 = 7	4 + 5 = 9
4 + 5 = 9	1 + 4 = 5	5 + 3 = 8	3 + 2 = 5	4 + 1 = 5	2 + 0 = 2	0 + 2 = 2	5 + 3 = 8	2 + 4 = 6	3 + 5 = 8
3 + 5 = 8	4 + 4 = 8	1 + 3 = 4	5 + 2 = 7	3 + 1 = 4	4 + 0 = 4	2 + 2 = 4	0 + 3 = 3	5 + 4 = 9	2 + 5 = 7
2 + 5 = 7	3 + 4 = 7	3 + 3 = 6	1 + 2 = 3	5 + 1 = 6	3 + 0 = 3	4 + 2 = 6	2 + 3 = 5	0 + 4 = 4	5 + 5 = 10
5 + 5 = 10	2 + 4 = 6	4 + 3 = 7	4 + 2 = 6	1 + 1 = 2	5 + 0 = 5	3 + 2 = 5	4 + 3 = 7	2 + 4 = 6	0 + 5 = 5

The **Section Diagnostic Tests** are specially arranged, too. The arrangement helps to identify if there are still problems and which facts those problems are.

Practice Sheet 6s, page 15.

14	7	10	11	16
10	9	16	8	11
12	14	15	14	15
9	6	7	16	7
7	15	11	13	10
6	10	9	7	14
11	12	6	9	13
16	15	16	12	9
15	8	14	8	12
7	11	16	10	6
6	9	14	7	13
9	7	10	11	15
10	12	8	6	16
16	14	15	16	9
13	8	6	7	11
8	10	12	16	8
11	13	9	13	12
12	15	7	12	10

Practice Sheet 7s, page 16.

15	11	13	10	8	7	17	14	16
13	12	9	14	15	11	10	7	8
10	15	7	16	17	13	14	9	17
12	16	17	11	9	15	13	8	10
11	13	15	8	12	10	7	16	9
17	14	13	7	16	17	11	15	12
8	13	9	12	10	15	16	17	7
9	17	16	8	11	7	14	10	13
16	10	14	17	15	9	12	8	11

Practice Sheet 8s, page 17.

12	11	18	10	13
14	16	12	17	15
11	8	9	18	9
9	17	13	15	12
8	12	11	9	16
13	14	8	11	15
18	17	18	14	12
17	10	16	10	14
9	13	18	12	8
8	11	16	9	15
11	9	12	13	17
12	14	10	8	18
18	16	17	18	11
15	10	8	13	8
10	12	14	18	10
13	15	11	15	14
14	17	9	17	10
17	8	11	14	12

Practice Sheet 9s, page 18.

17	13	15	12	10	9	19	16	18
15	14	11	16	17	13	12	9	10
12	17	9	18	19	15	16	11	19
14	18	19	13	11	17	15	10	12
13	15	17	10	14	12	9	18	11
19	16	15	9	18	19	13	17	14
9	11	18	13	16	10	12	15	18
18	12	16	19	17	11	14	10	13
11	19	18	10	13	9	16	12	15

Practice Sheet 10s, page 1

18	11	14	15	20
14	13	20	12	15
16	18	19	18	19
13	10	11	20	11
11	19	15	17	14
10	14	13	11	18
15	16	10	13	17
20	19	20	16	13
19	12	18	12	16
11	15	20	14	10
10	13	18	11	17
13	11	14	15	19
14	16	12	10	20
20	18	19	20	13
17	12	10	11	15
12	14	16	20	12
15	17	13	17	16
16	19	11	16	14

Answers
Addition

Section Diagnostic Test 6–10, page 20.

6s	7s	8s	9s	10s	6s	7s	8s	9s	10s
2 + 6 = 8	9 + 7 = 16	1 + 8 = 9	7 + 9 = 16	3 + 10 = 13	9 + 6 = 15	3 + 7 = 10	5 + 8 = 13	8 + 9 = 17	1 + 10 = 11
4 + 6 = 10	8 + 7 = 15	3 + 8 = 11	6 + 9 = 15	5 + 10 = 15	8 + 6 = 14	4 + 7 = 11	7 + 8 = 15	4 + 9 = 13	9 + 10 = 19
6 + 6 = 12	10 + 7 = 17	9 + 8 = 17	5 + 9 = 14	7 + 10 = 17	7 + 6 = 13	0 + 7 = 7	2 + 8 = 10	5 + 9 = 14	8 + 10 = 18
8 + 6 = 14	6 + 7 = 13	0 + 8 = 8	4 + 9 = 13	9 + 10 = 19	6 + 6 = 12	5 + 7 = 12	4 + 8 = 12	6 + 9 = 15	10 + 10 = 20
10 + 6 = 16	5 + 7 = 12	6 + 8 = 14	3 + 9 = 12	10 + 10 = 20	5 + 6 = 11	10 + 7 = 17	8 + 8 = 16	7 + 9 = 16	2 + 10 = 12
1 + 6 = 7	5 + 7 = 12	10 + 8 = 18	2 + 9 = 11	8 + 10 = 18	4 + 6 = 10	9 + 7 = 16	9 + 8 = 17	8 + 9 = 17	3 + 10 = 13
3 + 6 = 9	4 + 7 = 11	7 + 8 = 15	1 + 9 = 10	6 + 10 = 16	10 + 6 = 16	8 + 7 = 15	10 + 8 = 18	9 + 9 = 18	0 + 10 = 10
5 + 6 = 11	2 + 7 = 9	5 + 8 = 13	0 + 9 = 9	4 + 10 = 14	9 + 6 = 15	7 + 7 = 14	6 + 8 = 14	10 + 9 = 19	8 + 10 = 18
7 + 6 = 13	1 + 7 = 8	8 + 8 = 16	10 + 9 = 19	2 + 10 = 12	0 + 6 = 6	6 + 7 = 13	3 + 8 = 11	2 + 9 = 11	9 + 10 = 19
4 + 6 = 10	3 + 7 = 10	1 + 8 = 9	2 + 9 = 11	6 + 10 = 16	7 + 6 = 13	8 + 7 = 15	10 + 8 = 18	9 + 9 = 18	0 + 10 = 10

Review Sheet 0s–10s, page 21.

15	10	15	10	13	7	7	5	8
9	14	11	14	8	8	8	5	9
14	7	13	12	13	15	11	9	8
0	6	14	16	13	12	9	12	5
16	6	12	11	10	14	11	16	9
10	2	12	11	8	9	15	10	17
10	8	4	18	7	13	8	16	9
10	11	13	10	20	5	6	7	17
19	18	12	3	8	14	4	12	6

Review Sheet 0s–10s, page 22.

5 + 6 = 11	4 + 4 = 8	3 + 9 = 12	2 + 7 = 9
6 + 1 = 7	5 + 7 = 12	4 + 3 = 7	3 + 8 = 11
3 + 7 = 10	6 + 2 = 8	5 + 8 = 13	4 + 2 = 6
8 + 8 = 16	7 + 4 = 11	6 + 3 = 9	5 + 9 = 14
9 + 8 = 17	8 + 9 = 17	7 + 5 = 12	6 + 4 = 10
9 + 9 = 18	9 + 10 = 19	8 + 1 = 9	7 + 6 = 13
9 + 5 = 14	9 + 3 = 12	9 + 6 = 15	8 + 2 = 10
8 + 4 = 12	4 + 9 = 13	7 + 2 = 9	9 + 4 = 13
8 + 6 = 14	8 + 9 = 17	9 + 8 = 17	10 + 5 = 15

9 + 6 = 15	1 + 9 = 10	8 + 7 = 15	3 + 7 = 10	9 + 4 = 13	4 + 3 = 7	5 + 2 = 7	4 + 1 = 5	8 + 0 = 8
4 + 5 = 9	7 + 8 = 15	2 + 9 = 11	6 + 8 = 14	1 + 7 = 8	4 + 4 = 8	5 + 3 = 8	3 + 2 = 5	8 + 1 = 9
10 + 4 = 14	2 + 5 = 7	7 + 6 = 13	3 + 9 = 12	5 + 8 = 13	6 + 7 = 13	7 + 4 = 11	6 + 3 = 9	6 + 2 = 8
0 + 0 = 0	2 + 4 = 6	9 + 5 = 14	10 + 6 = 16	4 + 9 = 13	4 + 8 = 12	2 + 7 = 9	8 + 4 = 12	2 + 3 = 5

Final Assessment Test, page 23.

1 + 9 = 10	0 + 8 = 8	5 + 7 = 12	2 + 4 = 6	8 + 3 = 11	0 + 2 = 2	3 + 1 = 4	7 + 0 = 7	3 + 0 = 3	9 + 10 = 19	10s
8 + 6 = 14	0 + 9 = 9	1 + 8 = 9	10 + 7 = 17	5 + 4 = 9	5 + 3 = 8	3 + 2 = 5	4 + 1 = 5	9 + 0 = 9	1 + 0 = 1	0s
7 + 5 = 12	9 + 6 = 15	9 + 9 = 18	2 + 8 = 10	9 + 7 = 16	4 + 4 = 8	4 + 3 = 7	5 + 2 = 7	8 + 1 = 9	8 + 0 = 8	0s
2 + 4 = 6	3 + 5 = 8	10 + 6 = 16	8 + 9 = 17	3 + 8 = 11	2 + 7 = 9	7 + 4 = 11	2 + 3 = 5	7 + 2 = 9	2 + 1 = 3	1s
0 + 0 = 0	10 + 4 = 14	9 + 5 = 14	7 + 6 = 13	7 + 9 = 16	4 + 8 = 12	8 + 7 = 15	8 + 4 = 12	6 + 3 = 9	6 + 2 = 8	2s
8 + 8 = 16	3 + 3 = 6	7 + 4 = 11	6 + 5 = 11	3 + 6 = 9	6 + 9 = 15	5 + 8 = 13	1 + 7 = 8	9 + 4 = 13	1 + 3 = 4	3s
0 + 3 = 3	1 + 1 = 2	6 + 6 = 12	8 + 4 = 12	2 + 5 = 7	4 + 6 = 10	5 + 9 = 14	6 + 8 = 14	3 + 7 = 10	5 + 4 = 9	4s
9 + 2 = 11	5 + 3 = 8	2 + 2 = 4	9 + 9 = 18	3 + 4 = 7	1 + 5 = 6	1 + 6 = 7	4 + 9 = 13	7 + 8 = 15	6 + 7 = 13	7s
0 + 10 = 10	8 + 2 = 10	10 + 3 = 13	5 + 5 = 10	10 + 10 = 20	1 + 4 = 5	8 + 5 = 13	2 + 6 = 8	3 + 9 = 12	9 + 8 = 17	8s
5 + 10 = 15	8 + 10 = 18	10 + 2 = 12	7 + 3 = 10	4 + 4 = 8	7 + 7 = 14	0 + 4 = 4	4 + 5 = 9	0 + 6 = 6	2 + 9 = 11	9s

10s 10s 2s 3s DOUBLES 4s 5s 6s

The **Final Assessment Test** is arranged like the Beginning Assessment Test with diagonal facts.

Answers
Subtraction

Beginning Assessment Test, page 26.

9 − 9 = 0	10 − 8 = 2	9 − 7 = 2	11 − 4 = 7	3 − 3 = 0	4 − 2 = 2	1 − 1 = 0	8 − 0 = 8	6 − 0 = 6	0 − 0 = 0	0s
8 − 6 = 2	10 − 9 = 1	9 − 8 = 1	8 − 7 = 1	12 − 4 = 8	7 − 3 = 4	3 − 2 = 1	2 − 1 = 1	9 − 0 = 9	5 − 0 = 5	0s
10 − 5 = 5	7 − 6 = 1	11 − 9 = 2	8 − 8 = 0	7 − 7 = 0	13 − 4 = 9	8 − 3 = 5	5 − 2 = 3	3 − 1 = 2	10 − 0 = 10	0s
16 − 7 = 9	9 − 5 = 4	6 − 6 = 0	12 − 9 = 3	11 − 8 = 3	11 − 7 = 4	14 − 4 = 10	9 − 3 = 6	6 − 2 = 4	4 − 1 = 3	1s
8 − 8 = 0	17 − 7 = 10	8 − 5 = 3	9 − 6 = 3	13 − 9 = 4	12 − 8 = 4	12 − 7 = 5	4 − 4 = 0	6 − 3 = 3	7 − 2 = 5	2s
5 − 4 = 1	16 − 8 = 8	7 − 7 = 0	7 − 5 = 2	11 − 6 = 5	14 − 9 = 5	13 − 8 = 5	13 − 7 = 6	5 − 4 = 1	8 − 3 = 5	3s
6 − 3 = 3	6 − 4 = 2	15 − 8 = 7	8 − 7 = 1	6 − 5 = 1	12 − 6 = 6	15 − 9 = 6	14 − 8 = 6	14 − 7 = 7	6 − 4 = 2	4s
7 − 2 = 5	5 − 3 = 2	7 − 4 = 3	14 − 8 = 6	9 − 7 = 2	5 − 5 = 0	13 − 6 = 7	16 − 9 = 7	15 − 8 = 7	15 − 7 = 8	7s
20 − 10 = 10	8 − 2 = 6	4 − 3 = 1	8 − 4 = 4	13 − 8 = 5	10 − 7 = 3	11 − 5 = 6	14 − 6 = 8	17 − 9 = 8	16 − 8 = 8	8s
18 − 10 = 8	15 − 10 = 5	9 − 2 = 7	10 − 3 = 7	9 − 4 = 5	9 − 8 = 1	11 − 7 = 4	12 − 5 = 7	15 − 6 = 9	18 − 9 = 9	9s
10s	10s	2s	3s	4s	8s	7s	5s	6s		

Practice Sheet 0s & 1s, page 27.

9	2	1	0	6	8	6	5	9
1	0	10	8	7	4	3	0	2
8	3	1	0	6	8	5	5	9
6	1	5	3	1	0	9	6	7
2	8	1	7	5	2	0	10	7
5	4	10	3	9	7	3	1	0
3	6	6	0	5	0	9	4	2
0	1	7	8	1	7	2	0	5
6	2	0	8	0	2	9	4	2

Practice Sheet 2s, page 28.

0	6	2	4	1
6	2	1	10	8
7	3	4	0	5
2	4	5	3	0
9	1	6	10	7
0	6	4	8	1
10	3	7	5	2
3	1	9	7	0
5	2	4	6	8
7	0	2	10	3
1	9	5	4	10
5	1	8	7	4
3	9	0	6	2
6	0	3	1	6
4	7	1	5	8
2	9	5	1	4
6	5	3	7	10
8	6	7	5	3

Practice Sheet 3s, page 29.

3	2	9	4	5	1	0	10	8
0	5	7	6	1	3	4	9	2
1	3	5	8	2	0	10	6	9
8	4	3	7	6	9	1	5	2
7	9	6	1	4	10	0	3	5
6	0	4	8	5	10	2	7	1
9	0	6	8	1	7	4	0	3
2	6	4	1	9	5	3	8	0
5	1	3	0	8	9	2	4	6

Practice Sheet 4s, page 30.

6	10	1	3	5
8	4	0	2	6
4	0	9	5	7
5	1	2	6	3
0	2	3	1	8
7	9	4	10	5
8	4	2	7	9
1	9	6	5	10
3	0	2	4	7
5	8	0	1	4
9	10	3	2	6
3	9	10	4	2
1	6	8	5	0
4	7	1	9	6
2	5	9	3	7
0	8	3	9	2
4	10	1	5	7
1	4	5	3	6

Practice Sheet 5s, page 31.

7	3	10	1	8	6	5	0	2
4	1	0	9	2	3	10	8	5
6	8	3	7	4	10	1	2	5
2	10	1	3	6	0	8	9	4
8	9	2	1	5	4	7	10	3
0	5	7	4	10	2	6	8	1
1	4	8	2	5	3	9	0	6
3	7	0	4	6	10	5	2	8
5	0	4	2	10	7	3	1	9

Answers
Subtraction

Section Diagnostic Test 0–5, page 32.

5 – 5 = 0	11 – 5 = 6	7 – 5 = 2	9 – 5 = 4	8 – 5 = 3	5s
9 – 4 = 5	6 – 4 = 2	10 – 4 = 6	13 – 4 = 9	7 – 4 = 3	4s
7 – 3 = 4	11 – 3 = 8	5 – 3 = 2	6 – 3 = 3	12 – 3 = 9	3s
10 – 2 = 8	2 – 2 = 0	5 – 2 = 3	8 – 2 = 6	11 – 2 = 9	2s
1 – 1 = 0	9 – 1 = 8	6 – 1 = 5	10 – 1 = 9	7 – 1 = 6	1s
0 – 0 = 0	2 – 0 = 2	10 – 0 = 10	7 – 0 = 7	6 – 0 = 6	0s
6 – 5 = 1	12 – 5 = 7	14 – 5 = 9	11 – 5 = 6	8 – 5 = 3	5s
12 – 4 = 8	9 – 4 = 5	7 – 4 = 3	14 – 4 = 10	12 – 4 = 8	4s
4 – 3 = 1	13 – 3 = 10	10 – 3 = 7	12 – 3 = 9	9 – 3 = 6	3s
9 – 2 = 7	7 – 2 = 5	4 – 2 = 2	6 – 2 = 4	10 – 2 = 8	2s
3 – 1 = 2	5 – 1 = 4	8 – 1 = 7	7 – 1 = 6	1 – 1 = 0	1s
5 – 0 = 5	1 – 0 = 1	9 – 0 = 9	8 – 0 = 8	0 – 0 = 0	0s
10 – 5 = 5	8 – 5 = 3	12 – 5 = 7	15 – 5 = 10	9 – 5 = 4	5s
8 – 4 = 4	5 – 4 = 1	7 – 4 = 3	10 – 4 = 6	6 – 4 = 2	4s
9 – 3 = 6	10 – 3 = 7	8 – 3 = 5	11 – 3 = 8	9 – 3 = 6	3s
8 – 2 = 6	3 – 2 = 1	10 – 2 = 8	12 – 2 = 10	2 – 2 = 0	2s
4 – 4 = 0	14 – 4 = 10	11 – 4 = 7	8 – 4 = 4	7 – 4 = 3	4s
12 – 5 = 7	9 – 5 = 4	10 – 5 = 5	15 – 5 = 10	13 – 5 = 8	5s
4 – 4 = 0	13 – 4 = 9	9 – 4 = 5	6 – 4 = 2	10 – 4 = 6	4s
6 – 3 = 3	4 – 3 = 1	12 – 3 = 9	13 – 3 = 10	3 – 3 = 0	3s

Practice Sheet 6s, page 33.

6	2	9	0	7	5	4	10	1
3	0	10	8	1	2	9	7	4
5	7	2	8	3	9	0	4	1
1	9	0	2	5	10	7	8	3
7	8	1	0	4	3	6	9	2
10	4	6	3	9	1	5	7	0
0	3	7	1	4	2	8	10	5
2	6	10	3	5	9	4	1	7
4	10	3	1	9	6	2	0	8

Practice Sheet 7s, page 34.

1	10	6	5	0
2	9	10	3	4
8	10	0	10	5
4	6	1	7	2
5	1	10	4	6
3	9	2	0	8
9	6	3	2	7
0	8	10	8	4
2	5	8	1	9
6	7	0	10	3
1	10	6	3	0
0	6	7	1	10
9	3	5	2	8
1	4	9	6	3
10	2	6	0	4
8	5	0	6	10
1	7	9	2	4
9	1	2	0	3

Practice Sheet 8s, page 35.

0	7	9	5	3	2	8	10	1
9	8	6	10	0	7	5	2	3
5	0	6	1	7	9	10	2	8
7	9	0	3	8	5	6	1	4
6	10	9	2	1	4	7	0	7
2	4	1	7	10	3	5	9	0
1	5	10	2	0	6	8	3	7
4	8	1	3	7	2	10	5	10
8	1	10	7	4	9	1	6	5

Practice Sheet 9s, page 36.

1	5	7	9	0
3	10	6	8	2
10	6	4	1	9
0	7	8	1	2
6	8	9	7	3
2	4	10	5	0
3	10	8	2	4
1	7	0	9	6
7	4	1	0	5
9	6	8	6	2
0	3	6	10	7
1	5	7	9	0
4	5	9	8	1
9	4	5	10	8
7	1	3	0	6
10	2	7	4	1
8	0	4	9	2
6	3	7	4	8

Practice Sheet 10s, page 37.

7	6	4	8	9	5	3	0	1
3	9	4	10	0	7	8	5	6
5	7	9	1	6	3	4	5	2
4	8	7	0	10	2	5	9	5
0	2	10	5	8	1	3	7	9
10	3	8	0	9	4	6	1	5
2	6	10	1	5	0	8	3	8
6	10	8	5	2	7	10	4	3
1	7	2	6	3	8	7	8	0

Answers
Subtraction

Section Diagnostic Test 6–10, page 38.

6s	7s	8s	9s	10s
9 − 6 = 3	7 − 7 = 0	18 − 8 = 10	15 − 9 = 6	15 − 10 = 5
10 − 6 = 4	17 − 7 = 10	12 − 8 = 4	17 − 9 = 8	10 − 10 = 0
12 − 6 = 6	8 − 7 = 1	17 − 8 = 9	16 − 9 = 7	17 − 10 = 7
9 − 6 = 3	16 − 7 = 9	11 − 8 = 3	12 − 9 = 3	12 − 10 = 2
11 − 6 = 5	15 − 7 = 8	10 − 8 = 2	9 − 9 = 0	18 − 10 = 8
7 − 6 = 1	10 − 7 = 3	16 − 8 = 8	10 − 9 = 1	14 − 10 = 4
10 − 6 = 4	14 − 7 = 7	9 − 8 = 1	13 − 9 = 4	19 − 10 = 9
8 − 6 = 2	11 − 7 = 4	17 − 8 = 9	18 − 9 = 9	13 − 10 = 3
13 − 6 = 7	13 − 7 = 6	8 − 8 = 0	14 − 9 = 5	11 − 10 = 1
15 − 6 = 9	9 − 7 = 2	15 − 8 = 7	11 − 9 = 2	16 − 10 = 6
12 − 6 = 6	12 − 7 = 5	11 − 8 = 3	19 − 9 = 10	20 − 10 = 10
6 − 6 = 0	7 − 7 = 0	13 − 8 = 5	9 − 9 = 0	12 − 10 = 2
10 − 6 = 4	8 − 7 = 1	15 − 8 = 7	13 − 9 = 4	15 − 10 = 5
14 − 6 = 8	16 − 7 = 9	10 − 8 = 2	15 − 9 = 6	14 − 10 = 4
11 − 6 = 5	9 − 7 = 2	14 − 8 = 6	17 − 9 = 8	11 − 10 = 1
16 − 6 = 10	10 − 7 = 3	16 − 8 = 8	12 − 9 = 3	16 − 10 = 6
9 − 6 = 3	15 − 7 = 8	8 − 8 = 0	14 − 9 = 5	17 − 10 = 7
15 − 6 = 9	13 − 7 = 6	18 − 8 = 10	11 − 9 = 2	13 − 10 = 3
13 − 6 = 7	11 − 7 = 4	12 − 8 = 4	10 − 9 = 1	10 − 10 = 0
16 − 6 = 10	14 − 7 = 7	9 − 8 = 1	16 − 9 = 7	18 − 10 = 8

Review Sheet 0s–10s, page 39.

10	9	1	9	6	9	2	9	0
7	8	3	7	4	3	2	8	7
4	5	4	7	3	8	3	2	6
8	5	2	9	1	6	5	9	4
8	9	8	9	10	8	0	7	3
2	8	6	9	1	10	8	5	6
7	4	8	8	3	7	10	4	9
3	5	5	7	8	9	9	6	7
0	7	4	8	1	8	9	9	10

Review Sheet 0s–10s, page 40.

16 − 8 = 8	7 − 6 = 1	11 − 9 = 2	8 − 8 = 0
13 − 4 = 9	8 − 3 = 5	5 − 2 = 3	10 − 0 = 10
16 − 7 = 9	9 − 5 = 4	12 − 9 = 3	11 − 8 = 3
11 − 7 = 4	10 − 4 = 10	9 − 3 = 6	6 − 2 = 4
17 − 7 = 10	8 − 5 = 3	9 − 6 = 3	13 − 9 = 4
13 − 8 = 5	12 − 7 = 5	6 − 3 = 3	7 − 2 = 5
16 − 8 = 8	7 − 7 = 0	7 − 5 = 2	11 − 6 = 5
13 − 8 = 5	13 − 7 = 6	5 − 4 = 1	8 − 3 = 5
14 − 9 = 5	13 − 7 = 6	15 − 7 = 8	8 − 7 = 1

6 − 5 = 1	12 − 6 = 6	15 − 9 = 6	14 − 8 = 6	14 − 7 = 7	6 − 4 = 2	7 − 2 = 5	5 − 3 = 2	7 − 4 = 3
14 − 8 = 6	13 − 6 = 7	16 − 9 = 7	15 − 8 = 7	15 − 7 = 8	20 − 10 = 10	8 − 2 = 6	13 − 8 = 5	10 − 7 = 3
11 − 5 = 6	6 − 4 = 2	17 − 9 = 8	16 − 8 = 8	10 − 3 = 7	18 − 10 = 8	15 − 10 = 5	9 − 4 = 5	11 − 7 = 4
0 − 0 = 0	19 − 10 = 9	17 − 8 = 9	13 − 9 = 4	15 − 7 = 8	11 − 10 = 1	17 − 11 = 6	19 − 9 = 10	9 − 6 = 3

Final Assessment Test, page 41.

16 − 9 = 7	10 − 8 = 2	9 − 7 = 2	8 − 4 = 4	6 − 3 = 3	6 − 2 = 4	4 − 1 = 3	0 − 0 = 0	5 − 0 = 5	8 − 0 = 8	0s
14 − 6 = 8	11 − 9 = 2	9 − 8 = 1	11 − 7 = 4	9 − 4 = 5	7 − 3 = 4	8 − 2 = 6	5 − 1 = 4	1 − 0 = 1	6 − 0 = 6	0s
5 − 5 = 0	15 − 6 = 9	12 − 9 = 3	17 − 8 = 9	12 − 7 = 5	10 − 4 = 6	8 − 3 = 5	7 − 2 = 5	6 − 1 = 5	2 − 0 = 2	0s
15 − 7 = 8	6 − 5 = 1	14 − 6 = 8	13 − 9 = 4	16 − 8 = 8	13 − 7 = 6	11 − 4 = 7	9 − 3 = 6	9 − 2 = 7	7 − 1 = 6	1s
18 − 8 = 10	13 − 7 = 6	7 − 5 = 2	13 − 6 = 7	14 − 9 = 5	15 − 8 = 7	14 − 7 = 7	12 − 4 = 8	11 − 3 = 8	10 − 2 = 8	2s
4 − 4 = 0	14 − 8 = 6	14 − 7 = 7	8 − 5 = 3	12 − 6 = 6	15 − 9 = 6	14 − 8 = 6	15 − 7 = 8	13 − 4 = 9	12 − 3 = 9	3s
13 − 3 = 10	5 − 4 = 1	15 − 8 = 7	15 − 7 = 8	9 − 5 = 4	11 − 6 = 5	16 − 9 = 7	13 − 8 = 5	16 − 7 = 9	14 − 4 = 10	4s
2 − 2 = 0	4 − 3 = 1	6 − 4 = 2	16 − 8 = 8	16 − 7 = 9	14 − 5 = 9	10 − 6 = 4	17 − 9 = 8	12 − 8 = 4	17 − 7 = 10	7s
13 − 10 = 3	2 − 2 = 0	5 − 3 = 2	7 − 4 = 3	17 − 8 = 9	7 − 7 = 0	13 − 5 = 8	9 − 6 = 3	18 − 9 = 9	11 − 8 = 3	8s
11 − 10 = 1	17 − 10 = 7	4 − 2 = 2	6 − 3 = 3	8 − 4 = 4	9 − 8 = 1	8 − 7 = 1	12 − 5 = 7	8 − 6 = 2	19 − 9 = 10	9s

10s 10s 2s 3s 4s 8s 7s 5s 6s

Answers
Multiplication

Beginning Assessment Test, page 44.

0 × 9 = 0	9 × 8 = 72	4 × 7 = 28	8 × 5 = 40	8 × 6 = 48	5 × 2 = 10	5 × 1 = 5	9 × 0 = 0	3 × 0 = 0	4 × 10 = 40	10s
1 × 6 = 6	1 × 9 = 9	8 × 8 = 64	9 × 7 = 63	5 × 5 = 25	4 × 6 = 24	8 × 2 = 16	2 × 1 = 2	5 × 0 = 0	8 × 0 = 0	0s
3 × 4 = 12	7 × 3 = 21	2 × 9 = 18	0 × 8 = 0	3 × 7 = 21	6 × 5 = 30	9 × 6 = 54	7 × 2 = 14	3 × 1 = 3	6 × 0 = 0	0s
9 × 2 = 18	5 × 4 = 20	3 × 3 = 9	3 × 9 = 27	3 × 8 = 24	7 × 7 = 49	4 × 5 = 20	5 × 6 = 30	6 × 2 = 12	4 × 1 = 4	1s
2 × 2 = 4	4 × 2 = 8	8 × 4 = 32	4 × 3 = 12	4 × 9 = 36	7 × 8 = 56	2 × 7 = 14	3 × 5 = 15	7 × 6 = 42	9 × 2 = 18	2s
8 × 8 = 64	7 × 7 = 49	0 × 2 = 0	4 × 4 = 16	6 × 3 = 18	5 × 9 = 45	2 × 8 = 16	5 × 7 = 35	9 × 5 = 45	3 × 6 = 18	6s
9 × 1 = 9	1 × 1 = 1	3 × 3 = 9	3 × 2 = 6	9 × 4 = 36	5 × 3 = 15	6 × 9 = 54	4 × 8 = 32	6 × 7 = 42	7 × 5 = 35	5s
6 × 10 = 60	6 × 1 = 6	5 × 5 = 25	6 × 6 = 36	6 × 2 = 12	6 × 4 = 24	2 × 3 = 6	7 × 9 = 63	6 × 8 = 48	8 × 7 = 56	7s
7 × 10 = 70	0 × 10 = 0	7 × 1 = 7	9 × 9 = 81	4 × 4 = 16	7 × 2 = 14	2 × 4 = 8	9 × 3 = 27	8 × 9 = 72	5 × 8 = 40	8s
2 × 10 = 20	5 × 10 = 50	9 × 10 = 90	8 × 1 = 8	0 × 0 = 0	5 × 5 = 25	5 × 2 = 10	7 × 4 = 28	8 × 3 = 24	9 × 9 = 81	9s
10s	10s	10s	1s	DOUBLES		2s	4s	3s		

Practice Sheet 0s & 1s, page 45.

3	0	0	9	0	0	7	0	5
0	8	2	0	1	0	6	0	4
7	0	0	0	9	0	5	0	0
0	8	0	2	1	0	4	0	6
0	0	0	0	0	9	3	0	7
0	8	2	0	0	6	4	5	0
0	9	0	0	3	0	7	0	2
8	0	1	0	0	0	5	0	6
2	0	3	0	9	0	0	7	0

Practice Sheet 2s, page 46.

4	10	8	12	6
2	0	10	14	18
18	12	16	4	14
16	8	6	10	2
2	6	16	0	18
8	12	14	4	10
0	10	18	14	8
4	16	12	2	6
14	8	10	16	2
10	6	0	18	4
6	2	14	12	0
18	4	8	16	10
10	0	6	18	8
14	2	4	12	16
12	18	6	14	0
16	4	10	2	8
10	18	16	18	14
8	0	14	6	12

Practice Sheet 3s, page 47.

6	24	12	18	9	3	0	15	21
27	18	15	6	21	24	12	9	0
3	9	24	0	27	12	18	21	6
21	12	18	24	3	15	9	0	27
9	3	21	18	0	27	6	12	24
15	0	6	27	12	21	3	9	18
18	27	9	21	0	24	6	15	3
24	6	15	27	3	12	0	21	9
0	15	27	21	12	6	24	18	3

Practice Sheet 4s, page 48.

32	16	12	0	4
36	20	24	8	28
16	24	28	20	8
4	12	32	0	36
8	32	24	4	12
0	20	36	28	16
20	12	0	36	8
28	16	24	4	32
36	8	16	32	20
12	4	28	24	0
28	12	4	32	24
20	0	8	36	16
32	8	20	12	4
24	36	12	28	0
16	0	28	12	24
32	8	20	36	4
20	32	36	28	0
24	4	16	24	12

Practice Sheet 5s, page 49.

45	30	25	10	35	40	20	15	0
20	5	30	10	25	15	40	45	35
5	15	40	0	45	20	30	35	10
35	20	30	40	5	25	15	0	45
15	5	35	30	0	45	10	20	40
25	0	10	45	20	35	5	15	30
30	45	15	35	0	40	10	25	5
40	10	25	45	5	20	0	35	15
0	25	45	35	20	10	40	30	5

Answers
Multiplication

Section Diagnostic Test 0–5, page 50.

4 × 5 20	5 × 4 20	5 × 3 15	7 × 2 14	8 × 1 8	5 × 0 0	6 × 2 12	3 × 3 9	3 × 4 12	3 × 5 15
5 × 5 25	6 × 4 24	9 × 3 27	2 × 2 4	5 × 1 5	4 × 0 0	1 × 2 2	0 × 3 0	6 × 4 24	5 × 5 25
7 × 5 35	9 × 4 36	2 × 3 6	0 × 2 0	3 × 1 3	6 × 0 0	9 × 2 18	5 × 3 15	7 × 4 28	7 × 5 35
9 × 5 45	8 × 4 32	1 × 3 3	5 × 2 10	3 × 1 3	7 × 0 0	8 × 2 16	1 × 3 3	3 × 4 12	9 × 5 45
8 × 5 40	0 × 4 0	3 × 3 9	4 × 2 8	4 × 1 4	7 × 0 0	3 × 2 6	8 × 3 24	0 × 4 0	1 × 5 5
2 × 5 10	1 × 4 4	5 × 3 15	2 × 2 4	6 × 1 6	5 × 0 0	3 × 2 6	5 × 3 15	8 × 4 32	9 × 5 45
4 × 5 20	9 × 4 36	7 × 3 21	4 × 2 8	9 × 1 9	5 × 0 0	5 × 2 10	8 × 3 24	4 × 4 16	7 × 5 35
6 × 5 30	3 × 4 12	9 × 3 27	8 × 2 16	6 × 1 6	6 × 0 0	7 × 2 14	6 × 3 18	5 × 4 20	8 × 5 40
8 × 5 40	5 × 4 20	3 × 3 9	0 × 2 0	4 × 1 4	7 × 0 0	6 × 2 12	6 × 3 18	7 × 4 28	5 × 5 25
0 × 5 0	2 × 4 8	0 × 3 0	9 × 2 18	0 × 1 0	5 × 0 0	1 × 2 2	9 × 3 27	6 × 4 24	1 × 5 5
5s	4s	3s	2s	1s	0s	2s	3s	4s	5s

Practice Sheet 6s, page 51.

48	24	18	0	6
54	30	36	12	42
24	36	42	30	12
6	18	48	0	54
12	48	36	6	18
0	30	54	28	24
30	18	0	54	12
42	24	36	48	6
54	12	24	18	30
18	6	42	36	0
42	18	6	48	36
30	0	12	54	24
48	12	30	18	6
36	54	18	42	0
24	0	42	18	36
48	12	30	54	24
30	48	54	42	0
24	6	36	18	48

Practice Sheet 7s, page 52.

63	42	35	14	49	56	28	21	0
28	7	42	14	35	21	56	63	49
7	21	56	0	63	28	42	49	14
49	28	42	56	7	35	21	0	63
21	7	49	42	0	63	14	28	56
35	0	14	63	28	49	7	21	42
42	63	21	49	0	56	14	35	7
56	14	35	63	7	28	0	49	21
0	35	63	49	28	14	56	42	7

Practice Sheet 8s, page 53.

64	32	24	0	8
72	40	48	16	56
32	48	56	40	16
8	24	64	0	72
16	64	48	8	24
0	40	72	56	32
40	24	0	72	16
56	32	48	64	8
72	16	64	32	40
24	8	56	48	0
56	24	8	64	48
40	0	16	72	32
64	16	40	24	8
48	72	24	56	0
32	0	56	24	48
64	16	40	72	8
40	64	72	56	0
32	8	48	24	48

Practice Sheet 9s, page 54.

81	54	45	18	63	72	36	27	0
36	9	54	18	45	27	72	81	63
9	27	72	0	81	36	54	63	18
63	36	54	72	9	45	27	0	81
27	9	63	54	0	81	18	36	72
45	0	18	81	36	63	9	27	54
54	81	27	63	0	72	18	45	9
72	18	45	81	9	36	0	63	27
0	45	81	63	36	18	72	54	9

Practice Sheet 10s, page 55.

80	40	30	0	10
90	50	60	20	70
40	60	70	50	20
10	30	80	0	90
20	80	60	10	30
0	50	90	70	40
50	30	0	90	20
70	40	60	30	10
90	20	40	80	50
30	10	70	60	0
70	30	10	80	60
50	0	20	90	40
80	20	50	30	10
60	90	30	70	0
40	0	70	30	60
80	20	50	90	10
50	80	90	70	0
40	10	60	30	60

Answers
Multiplication

Section Diagnostic Test 6–10, page 56.

7 × 6 42	5 × 7 35	5 × 8 40	1 × 9 9	0 ×10 0	5 × 6 30	6 × 7 42	7 × 8 56	4 × 9 36	3 ×10 30
9 × 6 54	1 × 7 7	8 × 8 64	9 × 9 81	2 ×10 20	2 × 6 12	8 × 7 56	1 × 8 8	8 × 9 72	0 ×10 0
8 × 6 48	3 × 7 21	6 × 8 48	7 × 9 63	5 ×10 50	0 × 6 0	6 × 7 42	0 × 8 0	9 × 9 81	7 ×10 70
5 × 6 30	2 × 7 14	4 × 8 32	5 × 9 45	6 ×10 60	6 × 6 36	4 × 7 28	3 × 8 24	7 × 9 63	5 ×10 50
1 × 6 6	4 × 7 28	2 × 8 16	3 × 9 27	4 ×10 40	4 × 6 24	2 × 7 14	5 × 8 40	6 × 9 54	2 ×10 20
0 × 6 0	0 × 7 0	0 × 8 0	2 × 9 18	8 ×10 80	8 × 6 48	0 × 7 0	6 × 8 48	7 × 9 63	4 ×10 40
4 × 6 24	5 × 7 35	1 × 8 8	4 × 9 36	9 ×10 90	6 × 6 36	7 × 7 49	9 × 8 72	4 × 9 36	8 ×10 80
3 × 6 18	7 × 7 49	3 × 8 24	6 × 9 54	7 ×10 70	2 × 6 12	3 × 7 21	8 × 8 64	6 × 9 54	6 ×10 60
2 × 6 12	9 × 7 63	5 × 8 40	8 × 9 72	4 ×10 40	9 × 6 54	6 × 7 42	4 × 8 32	9 × 9 81	3 ×10 30
5 × 6 30	6 × 7 42	7 × 8 56	0 × 9 0	1 ×10 10	0 × 6 0	5 × 7 35	2 × 8 16	5 × 9 45	9 ×10 90
6s	7s	8s	9s	10s	6s	7s	8s	9s	10s

Review Sheet 0s–10s, page 57.

14	6	24	15	0
54	63	36	14	4
42	15	45	27	20
64	24	12	35	18
42	0	0	48	42
81	48	21	20	32
32	5	45	24	4
8	16	56	48	50
63	12	27	0	8
4	54	81	70	72
3	16	48	32	25
36	12	21	4	18
12	72	54	28	16
18	56	0	18	40
49	10	28	36	72
35	63	40	42	9
30	64	6	49	36
54	90	72	12	4

Review Sheet 0s–10s, page 58.

7 × 5 35	7 × 6 42	6 × 8 48	3 × 8 24	7 × 7 49	6 × 5 30	7 × 2 14	6 × 3 18	3 × 7 21
4 × 8 32	5 × 8 40	4 × 9 36	9 × 8 72	3 ×10 30	10 × 9 90	3 × 2 6	5 × 5 25	4 × 7 28
10 × 6 60	6 × 4 24	6 × 9 54	2 × 4 8	6 × 3 18	5 × 4 20	10 × 0 0	9 × 7 63	3 × 4 12
3 × 3 9	1 ×10 10	6 × 6 36	5 × 9 45	9 × 4 36	7 ×10 70	9 × 8 72	7 × 4 28	9 × 3 27

5 × 5 = 25	7 × 6 = 42	10 × 4 = 40	0 × 8 = 0
3 × 4 = 12	8 × 3 = 24	8 × 2 = 16	6 × 8 = 48
9 × 7 = 63	9 × 10 = 90	7 × 7 = 49	8 × 8 = 64
4 × 7 = 28	8 × 4 = 32	9 × 9 = 81	6 × 2 = 12
2 × 5 = 10	8 × 5 = 40	9 × 6 = 54	2 × 2 = 4
2 × 8 = 16	7 × 3 = 21	6 × 3 = 18	9 × 2 = 18
2 × 7 = 14	0 × 7 = 0	2 × 10 = 20	5 × 6 = 30
3 × 5 = 15	8 × 7 = 56	1 × 1 = 1	8 × 9 = 72
5 × 9 = 45	6 × 6 = 36	10 × 7 = 70	2 × 7 = 14

Final Assessment Test, page 59.

0 × 9 0	5 × 8 40	2 × 7 14	6 × 5 30	9 × 6 54	5 × 2 10	3 × 1 3	3 × 0 0	7 × 0 0	8 ×10 80	10s
9 × 3 27	1 × 9 9	8 × 8 64	9 × 7 63	3 × 5 15	7 × 6 42	8 × 2 16	2 × 1 2	5 × 0 0	8 × 0 0	0s
3 × 4 12	8 × 3 24	2 × 9 18	0 × 8 0	3 × 7 21	7 × 5 35	3 × 6 18	7 × 2 14	5 × 1 5	6 × 0 0	0s
9 × 2 18	5 × 4 20	7 × 3 21	3 × 9 27	3 × 8 24	8 × 7 56	9 × 5 45	5 × 6 30	6 × 2 12	4 × 1 4	1s
2 × 2 4	4 × 2 8	4 × 4 16	6 × 3 18	4 × 9 36	7 × 8 56	4 × 7 28	4 × 5 20	8 × 6 48	9 × 2 18	2s
8 × 8 64	3 × 3 9	8 × 2 16	2 × 4 8	5 × 3 15	5 × 9 45	4 × 8 32	5 × 7 35	8 × 5 40	4 × 6 24	6s
9 × 1 9	1 × 1 1	7 × 7 49	3 × 2 6	9 × 4 36	4 × 3 12	6 × 9 54	4 × 8 32	6 × 7 42	5 × 5 25	5s
0 ×10 0	6 × 1 6	5 × 5 25	6 × 6 36	0 × 2 0	7 × 4 28	3 × 3 9	7 × 9 63	6 × 8 48	7 × 7 49	7s
2 ×10 20	6 ×10 60	7 × 1 7	9 × 9 81	0 × 0 0	6 × 2 12	8 × 4 32	2 × 3 6	8 × 9 72	9 × 8 72	8s
5 ×10 50	7 ×10 70	9 ×10 90	8 × 1 8	4 × 4 16	5 × 5 25	7 × 2 14	6 × 4 24	1 × 3 3	9 × 9 81	9s

10s 10s 10s 1s Doubles 2s 4s 3s

Answers
Division

Beginning Assessment Test, page 62.

$\overset{1}{3\overline{)3}}$	$\overset{2}{2\overline{)4}}$	$\overset{3}{1\overline{)3}}$	$\overset{2}{4\overline{)8}}$	$\overset{4}{3\overline{)12}}$	$\overset{4}{2\overline{)8}}$	$\overset{2}{1\overline{)2}}$	$\overset{2}{8\overline{)16}}$	$\overset{2}{9\overline{)18}}$	$\overset{1}{10\overline{)10}}$	10s
$\overset{1}{4\overline{)4}}$	$\overset{2}{3\overline{)6}}$	$\overset{1}{2\overline{)2}}$	$\overset{2}{1\overline{)2}}$	$\overset{1}{4\overline{)4}}$	$\overset{1}{3\overline{)3}}$	$\overset{6}{2\overline{)12}}$	$\overset{10}{1\overline{)10}}$	$\overset{7}{8\overline{)56}}$	$\overset{3}{9\overline{)27}}$	9s
$\overset{8}{5\overline{)40}}$	$\overset{3}{4\overline{)12}}$	$\overset{3}{3\overline{)9}}$	$\overset{3}{2\overline{)6}}$	$\overset{4}{1\overline{)4}}$	$\overset{3}{4\overline{)12}}$	$\overset{2}{3\overline{)6}}$	$\overset{1}{2\overline{)2}}$	$\overset{1}{1\overline{)1}}$	$\overset{8}{8\overline{)64}}$	8s
$\overset{2}{6\overline{)12}}$	$\overset{1}{5\overline{)5}}$	$\overset{2}{4\overline{)8}}$	$\overset{4}{3\overline{)12}}$	$\overset{5}{2\overline{)10}}$	$\overset{5}{1\overline{)5}}$	$\overset{5}{4\overline{)20}}$	$\overset{3}{3\overline{)9}}$	$\overset{2}{2\overline{)4}}$	$\overset{3}{1\overline{)3}}$	1s
$\overset{1}{7\overline{)7}}$	$\overset{1}{6\overline{)6}}$	$\overset{2}{5\overline{)10}}$	$\overset{9}{4\overline{)36}}$	$\overset{5}{3\overline{)15}}$	$\overset{7}{2\overline{)14}}$	$\overset{6}{1\overline{)6}}$	$\overset{10}{4\overline{)40}}$	$\overset{6}{3\overline{)18}}$	$\overset{3}{2\overline{)6}}$	2s
$\overset{2}{8\overline{)16}}$	$\overset{2}{7\overline{)14}}$	$\overset{3}{6\overline{)18}}$	$\overset{3}{5\overline{)15}}$	$\overset{4}{4\overline{)16}}$	$\overset{1}{3\overline{)3}}$	$\overset{8}{2\overline{)16}}$	$\overset{7}{1\overline{)7}}$	$\overset{4}{4\overline{)16}}$	$\overset{10}{3\overline{)30}}$	3s
$\overset{9}{9\overline{)81}}$	$\overset{1}{8\overline{)8}}$	$\overset{3}{7\overline{)21}}$	$\overset{8}{6\overline{)48}}$	$\overset{4}{5\overline{)20}}$	$\overset{5}{4\overline{)20}}$	$\overset{7}{3\overline{)21}}$	$\overset{9}{2\overline{)18}}$	$\overset{8}{1\overline{)8}}$	$\overset{6}{4\overline{)24}}$	4s
$\overset{6}{2\overline{)12}}$	$\overset{4}{9\overline{)36}}$	$\overset{3}{8\overline{)24}}$	$\overset{4}{7\overline{)28}}$	$\overset{4}{6\overline{)24}}$	$\overset{5}{5\overline{)25}}$	$\overset{6}{4\overline{)24}}$	$\overset{8}{3\overline{)24}}$	$\overset{10}{2\overline{)20}}$	$\overset{9}{1\overline{)9}}$	1s
$\overset{3}{10\overline{)30}}$	$\overset{4}{2\overline{)8}}$	$\overset{5}{9\overline{)45}}$	$\overset{4}{8\overline{)32}}$	$\overset{7}{7\overline{)49}}$	$\overset{6}{6\overline{)36}}$	$\overset{6}{5\overline{)30}}$	$\overset{7}{4\overline{)28}}$	$\overset{9}{3\overline{)27}}$	$\overset{6}{2\overline{)12}}$	2s
$\overset{2}{10\overline{)20}}$	$\overset{4}{10\overline{)40}}$	$\overset{9}{2\overline{)18}}$	$\overset{6}{9\overline{)54}}$	$\overset{5}{8\overline{)40}}$	$\overset{6}{7\overline{)42}}$	$\overset{7}{6\overline{)42}}$	$\overset{7}{5\overline{)35}}$	$\overset{8}{4\overline{)32}}$	$\overset{6}{3\overline{)18}}$	3s
10s	10s	2s	9s	8s	7s	6s	5s	4s		

Practice Sheet 1s & 2s, page 63.

1	1	2	2	3	3	4	4	5
6	4	7	7	8	1	9	9	10
9	1	6	8	6	7	10	10	9
10	8	3	4	5	5	1	3	7
2	6	6	1	10	5	4	2	6
10	7	9	8	9	7	10	6	10
10	1	8	2	3	3	2	2	1
2	4	3	9	4	8	5	7	6
7	5	8	4	9	3	10	2	2

Practice Sheet 3s, page 64.

1	3	4	5	9
6	2	1	6	7
5	8	4	8	3
2	4	1	5	2
9	1	7	2	7
3	7	8	6	9
6	2	5	1	4
2	6	1	8	5
4	3	10	1	3
7	5	4	10	9
2	9	8	6	8
8	6	7	3	2
1	7	6	9	4
4	1	10	5	10
5	4	3	8	3
7	6	1	9	1
10	2	10	3	6
8	9	4	10	7

Practice Sheet 4s, page 65.

1	3	6	5	10	7	4	8	2
5	3	10	9	1	6	4	1	8
9	2	3	10	4	5	3	6	2
1	10	9	2	8	4	6	7	4
2	7	5	9	1	3	10	4	6
3	8	4	1	6	2	4	10	9
8	9	10	8	9	8	10	7	6
3	1	2	7	4	8	5	2	3
7	4	2	5	1	9	6	10	4

Practice Sheet 5s, page 66.

2	1	9	1	7
4	5	7	8	2
3	6	2	9	10
8	10	6	3	1
1	5	4	2	3
10	8	9	1	2
2	10	3	4	5
3	7	2	5	6
5	2	1	9	1
7	1	6	2	8
4	6	7	6	4
10	2	5	4	1
9	5	3	7	8
1	10	7	6	2
7	8	10	3	1
8	3	5	10	5
2	9	8	2	9
3	4	5	4	3

Section Diagnostic Test 1s–5s, page 67.

10 ÷ 5 = 2	5 ÷ 5 = 1	45 ÷ 5 = 9	5 ÷ 5 = 1	35 ÷ 5 = 7	5s
20 ÷ 4 = 5	24 ÷ 4 = 6	36 ÷ 4 = 9	16 ÷ 4 = 4	40 ÷ 4 = 10	4s
15 ÷ 3 = 5	27 ÷ 3 = 9	6 ÷ 3 = 2	3 ÷ 3 = 1	30 ÷ 3 = 10	3s
14 ÷ 2 = 7	4 ÷ 2 = 2	20 ÷ 2 = 10	10 ÷ 2 = 5	8 ÷ 2 = 4	2s
1 ÷ 1 = 1	4 ÷ 1 = 4	7 ÷ 1 = 7	2 ÷ 1 = 2	3 ÷ 1 = 3	1s
50 ÷ 5 = 10	40 ÷ 5 = 8	45 ÷ 5 = 9	5 ÷ 5 = 1	10 ÷ 5 = 2	5s
4 ÷ 4 = 1	36 ÷ 4 = 9	12 ÷ 4 = 3	20 ÷ 4 = 5	8 ÷ 4 = 2	4s
6 ÷ 3 = 2	15 ÷ 3 = 5	21 ÷ 3 = 7	27 ÷ 3 = 9	3 ÷ 3 = 1	3s
4 ÷ 2 = 2	8 ÷ 2 = 4	16 ÷ 2 = 8	20 ÷ 2 = 10	18 ÷ 2 = 9	2s
25 ÷ 5 = 5	5 ÷ 5 = 1	30 ÷ 5 = 6	35 ÷ 5 = 7	40 ÷ 5 = 8	5s
20 ÷ 4 = 5	28 ÷ 4 = 7	20 ÷ 4 = 5	16 ÷ 4 = 4	32 ÷ 4 = 8	4s
30 ÷ 3 = 10	27 ÷ 3 = 9	12 ÷ 3 = 4	18 ÷ 3 = 6	24 ÷ 3 = 8	3s
10 ÷ 2 = 5	12 ÷ 2 = 6	14 ÷ 2 = 7	8 ÷ 2 = 4	6 ÷ 2 = 3	2s
8 ÷ 1 = 8	6 ÷ 1 = 6	10 ÷ 1 = 10	4 ÷ 1 = 4	7 ÷ 1 = 7	1s
35 ÷ 5 = 7	40 ÷ 5 = 8	50 ÷ 5 = 10	15 ÷ 5 = 3	5 ÷ 5 = 1	5s
40 ÷ 4 = 10	8 ÷ 4 = 2	12 ÷ 4 = 3	24 ÷ 4 = 6	16 ÷ 4 = 4	4s
15 ÷ 3 = 5	21 ÷ 3 = 7	18 ÷ 3 = 6	3 ÷ 3 = 1	6 ÷ 3 = 2	3s
6 ÷ 2 = 3	20 ÷ 2 = 10	10 ÷ 2 = 5	14 ÷ 2 = 7	2 ÷ 2 = 1	2s

Answers
Division

Practice Sheet 6s, page 68.

1	3	2	5	9	4	10	6	7
3	6	8	1	5	9	7	2	10
1	5	2	9	1	8	4	7	3
5	2	9	1	8	4	10	3	6
4	1	7	2	10	8	3	1	5
10	8	9	3	6	7	5	4	2
8	4	1	9	2	10	6	7	3
6	7	3	4	10	9	8	5	1
2	9	5	6	7	3	1	8	4

Practice Sheet 7s, page 69.

1	5	10	6	8
3	9	5	7	4
4	3	1	10	6
7	5	9	8	2
2	6	10	4	9
1	2	5	9	7
8	7	6	5	2
5	4	3	10	1
6	8	2	3	6
9	1	4	1	3
2	3	9	7	4
6	9	10	2	8
7	1	5	10	5
3	7	6	2	4
5	10	1	8	3
6	1	3	4	5
4	6	1	2	7
1	2	7	5	3

Practice Sheet 8s, page 70.

1	3	2	4	6	9	10	7	8
3	7	5	2	9	6	1	6	4
4	8	1	6	5	2	7	10	3
5	10	8	4	1	3	6	2	7
2	3	10	9	5	8	7	1	5
10	4	1	8	9	6	5	2	3
8	5	3	6	5	9	1	10	2
1	9	4	1	7	2	3	9	3
9	6	3	10	2	1	4	3	5

Practice Sheet 9s, page 71.

1	4	3	9	8
2	7	1	8	9
5	9	6	3	10
3	4	10	1	2
7	2	8	4	5
4	3	9	10	1
5	1	7	8	6
10	5	4	2	3
8	6	2	5	8
2	5	3	7	2
10	1	6	4	1
9	10	1	7	8
5	8	6	2	4
2	1	10	9	5
9	4	7	3	8
1	8	9	2	7
3	5	2	10	4
4	10	3	9	5

Practice Sheet 10s, page 72.

1	9	5	8	3	7	10	2	4
5	2	6	1	4	9	3	10	7
2	10	7	3	8	5	9	1	4
8	3	2	7	9	6	4	10	1
6	2	4	9	10	8	1	5	3
10	4	9	6	7	2	3	8	5
5	1	8	10	2	9	4	6	9
4	3	2	1	5	10	6	9	7
9	8	10	6	7	1	2	3	4

Section Diagnostic Test 6s–10s, page 73.

42 ÷ 6 = 7	54 ÷ 6 = 9	48 ÷ 6 = 8	30 ÷ 6 = 5	36 ÷ 6 = 6	6s
7 ÷ 7 = 1	21 ÷ 7 = 3	14 ÷ 7 = 2	28 ÷ 7 = 4	70 ÷ 7 = 10	7s
24 ÷ 8 = 3	48 ÷ 8 = 6	64 ÷ 8 = 8	8 ÷ 8 = 1	32 ÷ 8 = 4	8s
81 ÷ 9 = 9	45 ÷ 9 = 5	9 ÷ 9 = 1	54 ÷ 9 = 6	27 ÷ 9 = 3	9s
6 ÷ 6 = 1	60 ÷ 6 = 10	24 ÷ 6 = 4	18 ÷ 6 = 3	12 ÷ 6 = 2	6s
35 ÷ 7 = 5	49 ÷ 7 = 7	63 ÷ 7 = 9	56 ÷ 7 = 8	42 ÷ 7 = 6	7s
80 ÷ 8 = 10	48 ÷ 8 = 6	16 ÷ 8 = 2	32 ÷ 8 = 4	64 ÷ 8 = 8	8s
27 ÷ 9 = 3	36 ÷ 9 = 4	18 ÷ 9 = 2	45 ÷ 9 = 5	63 ÷ 9 = 7	9s
90 ÷ 10 = 9	20 ÷ 10 = 2	50 ÷ 10 = 5	100 ÷ 10 = 10	70 ÷ 10 = 7	10s
72 ÷ 9 = 8	54 ÷ 9 = 6	27 ÷ 9 = 3	90 ÷ 9 = 10	72 ÷ 9 = 8	9s
32 ÷ 8 = 4	40 ÷ 8 = 5	32 ÷ 8 = 4	16 ÷ 8 = 2	8 ÷ 8 = 1	8s
63 ÷ 7 = 9	49 ÷ 7 = 7	35 ÷ 7 = 5	21 ÷ 7 = 3	7 ÷ 7 = 1	7s
18 ÷ 6 = 3	6 ÷ 6 = 1	54 ÷ 6 = 9	30 ÷ 6 = 5	42 ÷ 6 = 7	6s
10 ÷ 5 = 2	5 ÷ 5 = 1	45 ÷ 5 = 9	5 ÷ 5 = 1	35 ÷ 5 = 7	5s
12 ÷ 6 = 2	60÷ 6 = 10	30 ÷ 6 = 5	24 ÷ 6 = 4	48 ÷ 6 = 8	6s
14 ÷ 7 = 2	21 ÷ 7 = 3	35 ÷ 7 = 5	7 ÷ 7 = 1	70 ÷ 7 = 10	7s
64 ÷ 8 = 8	24 ÷ 8 = 3	40 ÷ 8 = 5	16 ÷ 8 = 2	8 ÷ 8 = 1	8s
81 ÷ 9 = 9	72 ÷ 9 = 8	45 ÷ 9 = 5	18 ÷ 9 = 2	27 ÷ 9 = 3	9s

Answers
Division

Review Sheet, page 74.

1	8	6	8	5	9	4	3	1
9	9	6	6	4	4	3	7	10
6	8	7	5	7	4	8	7	9
5	7	3	6	8	3	7	9	6
10	8	2	1	3	3	4	2	9
6	6	2	10	2	2	6	6	5
10	6	7	6	7	4	6	3	9
3	5	8	6	3	5	9	10	8
9	6	7	10	8	9	8	4	8

Review Sheet 0s–10s, page 75.

1 5)5	5 6)30	6 7)42	3 8)24	9 3)27	10 10)100	4 8)32	7 5)35	7 7)49
7 5)35	8 9)72	9 4)36	6 2)12	7 8)56	4 7)28	5 8)40	10 2)20	6 6)36
8 4)32	1 7)7	3 6)18	8 5)40	7 3)21	9 9)81	1 3)3	9 7)63	7 8)56
3 8)24	7 7)49	6 9)54	4 3)12	9 2)18	6 8)48	9 4)36	5 9)45	4 5)20

30 ÷ 5 = 6	42 ÷ 6 = 7	40 ÷ 10 = 4	24 ÷ 8 = 3
63 ÷ 9 = 7	32 ÷ 4 = 8	8 ÷ 2 = 4	18 ÷ 3 = 6
45 ÷ 5 = 9	90 ÷ 10 = 9	49 ÷ 7 = 7	64 ÷ 8 = 8
36 ÷ 9 = 4	12 ÷ 4 = 3	72 ÷ 9 = 8	30 ÷ 6 = 5
10 ÷ 5 = 2	50 ÷ 5 = 10	54 ÷ 6 = 9	16 ÷ 4 = 4
16 ÷ 8 = 2	21 ÷ 3 = 7	27 ÷ 3 = 9	18 ÷ 2 = 9
100 ÷ 10 = 10	20 ÷ 5 = 4	32 ÷ 8 = 4	27 ÷ 3 = 9
12 ÷ 6 = 2	56 ÷ 7 = 8	1 ÷ 1 = 1	30 ÷ 5 = 6
21 ÷ 3 = 7	14 ÷ 2 = 7	24 ÷ 4 = 6	32 ÷ 4 = 8

Final Assessment Test, page 76.

2 9)18	2 8)16	1 7)7	10 6)60	4 5)20	4 4)16	1 5)5	3 6)18	4 7)28	10 10)100	10s
4 3)12	6 9)54	6 8)48	5 7)35	2 6)12	6 5)30	7 4)28	2 5)10	6 6)36	3 7)21	7s
2 4)8	1 3)3	8 9)72	9 8)72	2 7)14	9 6)54	9 5)45	5 4)20	10 5)50	8 6)48	6s
3 2)6	3 4)12	3 3)9	3 9)27	8 8)64	9 7)63	5 6)30	3 5)15	9 4)36	8 5)40	5s
3 1)3	2 2)4	10 4)40	10 3)30	9 9)81	4 8)32	10 7)70	8 6)48	7 5)35	6 4)24	4s
9 3)27	4 1)4	1 2)2	8 4)32	5 3)15	4 9)36	1 8)8	6 7)42	4 6)24	2 5)10	5s
1 8)8	6 3)18	7 1)7	4 2)8	3 4)12	9 3)27	2 9)18	5 8)40	8 7)56	7 6)42	6s
9 2)18	7 8)56	10 3)30	5 1)5	5 2)10	9 4)36	6 3)18	1 9)9	10 8)80	7 7)49	7s
6 10)60	10 2)20	9 8)72	7 3)21	6 1)6	8 2)16	6 4)24	7 3)21	2 9)18	3 8)24	8s
8 10)80	3 10)30	7 2)14	6 8)48	2 3)6	8 1)8	6 2)12	7 4)28	3 3)9	10 9)90	9s

10s 10s 2s 8s 3s 1s 2s 4s 3s